DISTRICT COOLING

Theory and Practice

HEAT TRANSFER

A Series of Reference Books and Textbooks

SERIES EDITOR

Afshin J. Ghajar

Regents Professor
School of Mechanical and Aerospace Engineering
Oklahoma State University

District Cooling: Theory and Practice, *Alaa A. Olama*

Introduction to Compressible Fluid Flow, Second Edition, *Patrick H. Oosthuizen and William E. Carscallen*

Advances in Industrial Heat Transfer, *Alina Adriana Minea*

Introduction to Thermal and Fluid Engineering, *Allan D. Kraus, James R. Welty, and Abdul Aziz*

Thermal Measurements and Inverse Techniques, *Helcio R.B. Orlande, Olivier Fudym, Denis Maillet, and Renato M. Cotta*

Conjugate Problems in Convective Heat Transfer, *Abram S. Dorfman*

Engineering Heat Transfer: Third Edition, *William S. Janna*

DISTRICT **COOLING**

Theory and **Practice**

Alaa A. Olama

CRC Press
Taylor & Francis Group
Boca Raton London New York

CRC Press is an imprint of the
Taylor & Francis Group, an **informa** business

CRC Press
Taylor & Francis Group
6000 Broken Sound Parkway NW, Suite 300
Boca Raton, FL 33487-2742

Printed by CPI on sustainably sourced paper
Version Date: 20160826

International Standard Book Number-13: 978-1-4987-0550-9 (Hardback)

Library of Congress Cataloging-in-Publication Data

Names: Olama, Alaa A., author.
Title: District cooling : theory and practice / Alaa A. Olama.
Description: Boca Raton : CRCress, 2017. | Series: Heat transfer : a series of reference books and textbooks ; 7 | Includes bibliographical references and index.
Identifiers: LCCN 2016023722 | ISBN 9781498705509 (alk. paper)
Subjects: LCSH: Air conditioning from central stations.
Classification: LCC TH7687.75 O43 2017 | DDC 697.9/3--dc23
LC record available at https://lccn.loc.gov/2016023722

Visit the Taylor & Francis Web site at
http://www.taylorandfrancis.com

and the CRC Press Web site at
http://www.crcpress.com

Dedication

In Memory:
Knowledge builds a pillarless house while ignorance destroys a house of glory and honour.

A. Shawky (1868–1932)
To my father who engraved it into my soul.

Contents

List of Tables

List of Figures

Preface

I have been working in refrigeration and air-conditioning, both academically and in the industry, since 1971. In 2002, I came face to face with district cooling. This was when I was asked to conduct a study for a city, in a high-ambient temperature country, to adopt an air-conditioning strategy for the next 50 years. A major place of assembly was located in the city center. Chilled water from a dedicated remote district cooling plant supplied the air-conditioned systems for this place of assembly. Its capacity was about 105,000 kW (30,000 TR). Hotels, hostels, motels, and other lodging facilities, in total 320 such establishments, surround this major place of assembly. These air-cooled establishments were rejecting so much heat that a steady increase in ambient temperatures occurred in the city center and over the years, the comfort conditions for the whole city center progressively deteriorated.

An obvious solution to this problem was to connect all these buildings to a second district cooling plant and locate the plant in a remote area well away from the city center. But where should this station be located? How far away should it be from the center—5 or 10 or 15 km away? Would there be enough room in the underground utility tunnels that connected the buildings together to install supply and return chilled water piping? Would I need to decrease the size of the pipes, using perhaps ice-slurry solutions, to accommodate the unplanned chilled water piping in the tunnels?

All these issues made me realize that a heating, ventilation, and air-conditioning (HVAC) engineer is not necessarily qualified to answer these questions. I searched for references on district cooling; there were a few available on district heating, but when it came to district cooling, it was a different matter. Sometimes, if I was lucky, I would come across a spattering of information in the district heating references, which also addressed district cooling issues, but not extensively and certainly not for a high-ambient temperature country.

Slowly, a district cooling best practices guide appeared in 2008, together with information in the American Society of Heating, Air-Conditioning, and Refrigeration Engineers (ASHRAE) handbooks. There was, however, no district cooling reference book available, neither on the principles nor on the theory and practice.

In 2005, I became obsessed with forming a district cooling company and started exploring this field. Once I left the company, which I established in 2012 with four other partner companies, a close friend and colleague suggested that I write about district cooling. The resulting effort is this book. I hope you find it useful.

Alaa A. Olama
Independent Consultant

Author

Alaa A. Olama, Ph.D., has 35 years of experience in designing refrigeration and air-conditioning. He received both an M.Sc. and Ph.D. from King's College, London University, England, in mechanical engineering specializing in refrigeration and air-conditioning. He is a member of the Refrigeration, Air Conditioning and Heat Pumps Technical Options Committee (RTOC) formed by the Technical and Economical Assessment Panel of the United Nations Environment Programme (UNEP) to assess the development of relevant technologies to replace ozone-depleting substances (ODSs) in the fields of refrigeration and air-conditioning, under the Montréal Protocol. He is the past president of the board of directors of the American Society of Heating, Refrigerating, and Air-Conditioning Engineers (ASHRAE) Cairo Chapter, 2002–2003; and was general chair of the Second Annual Regional Conference (ARC) of ASHRAE's Region-at-Large held in Cairo in September 2003.

Dr. Olama is the founder, board of directors member, and vice chair of the first district cooling company in Egypt. He is the head of the committee writing the first District Cooling Code for Egypt. Recognized as an international expert in district cooling, he is also a member of the committee writing the Egyptian Code of Air-Conditioning, Refrigeration, and Automatic Control, and a member of a committee writing the Arab Refrigeration and Air-Conditioning Code. He is a member of the International Reviewer's Panel of PRAHA, formed by the United Nations Industrial Development Organization (UNIDO) and UNEP for testing new Low-Global Warming Potential (GWP) refrigerants in high-ambient temperature countries in the Gulf. He is the technical advisor for the Egyptian Low-GWP Refrigerants testing program in Egypt (EGYPRA) and a member of the expert panel of the Low-GWP Refrigerants testing program for high-ambient countries for the U.S. Department of Energy (DoE) at Oak Ridge National Laboratory (ORNL). Dr Olama is an independent consultant.

1 Introduction to District Cooling

1.1 DEFINING DISTRICT COOLING

A district cooling (DC) system is a central air-conditioning system that produces and distributes chilled water from a plant(s) to buildings, thus centralizing the production of chilled water and maximizing economy of scale.

District cooling systems produce and deliver chilled water, or a secondary fluid, from a central source to consumers in a more efficient, reliable, and environment-friendly way than in-building air-conditioning stations. Consumers may be residential, commercial, industrial, or other users in need of chiller water. Comfort air-conditioning or process cooling systems use chilled media to operate their systems. Figure 1.1 shows a district cooling system connected to consumers.

To maintain comfort conditions, individual room air-conditioners generate cooling energy locally for one room. Central cooling systems generate cooling energy in one or more central places within a building and distribute it to more than one room within the building.

District cooling systems generate cooling energy centrally and distribute it to various users' buildings by utilizing a piping network.

The capital cost of individual air-conditioners is normally low when compared to central air-conditioners, but its operating costs are high because their energy efficiency is low. Capital costs of central air-conditioners are higher, but so is their energy efficiency making their operating costs lower. Over their operating lifetime, the overall cost of central air-conditioners becomes lower than individual air-conditioners.

This analogy applies to a comparison between district cooling systems and central air-conditioners. District cooling systems have a higher capital cost than central air-conditioners and a higher energy efficiency. This is particularly so when buildings are situated in a dense area where cooling loads are high per surface area and have a diversified use. If a small number of rooms in a building need air-conditioning, it may be best to use individual air-conditioners; density of cooling load per surface area is usually a deciding factor.

Across the board, utilization of district cooling for all buildings is not usually a sound proposition because some applications may best be served with individual air-conditioners. For one or more building in a heavily populated area, district cooling is usually a good option.

Inside a building, the air-side equipment, such as air handlers, fan coil units, terminal units, and chilled water distribution systems, remain the same when a district cooling system supplies chilled water instead of a local in-building chilled water plant.

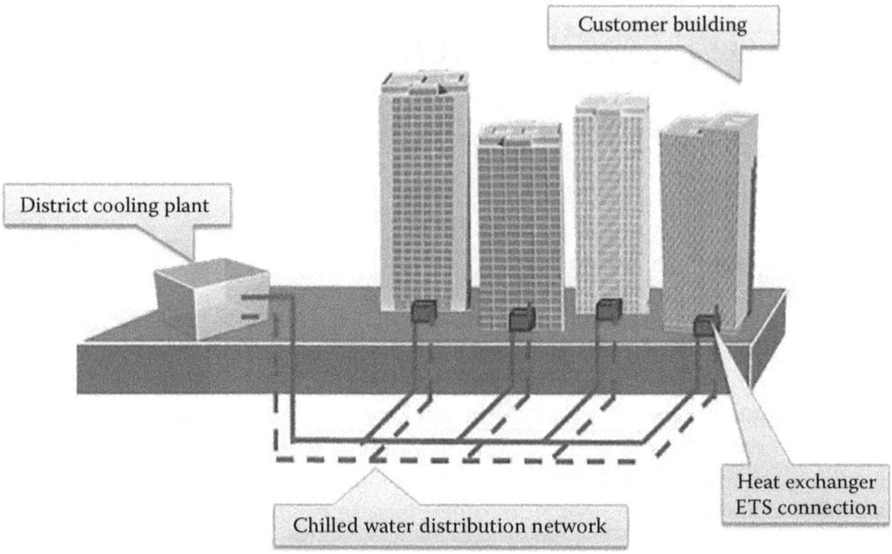

FIGURE 1.1 A district cooling system connected to end users.

New control systems, pumps, chilled water heat exchangers, and energy metering systems are some of the additions needed for district cooling connected buildings.

An important issue for a district cooling system is its higher energy efficiency. This is especially important in countries where energy supply is a factor. The reduction of carbon dioxide emissions of district cooling systems compared to individual or central air-conditioners is another important issue.

District cooling systems are experiencing a rapid growth in several developing countries, especially in the Middle East, where the largest growth in DC occurred during the past 10 years. The total installed air-conditioning capacity in the Gulf exceeds 8,800,000 kW (2,500,000 TR [tons of refrigeration]), making it the largest installed capacity in the world.[1]

1.2 THE ECONOMIC AND ENVIRONMENTAL BENEFITS OF DISTRICT COOLING FOR A CITY

The Sankey diagram in Figure 1.2 shows two scenarios to provide heating, cooling, and electricity to a city.[2] One scenario uses a traditional coal-fired power station, business as usual (BAU) scenario, whereas the second scenario uses natural gas in a modern combined heat and power (CHP) station.

In the first scenario with a conventional power station, the typical average thermal efficiency of this simple cycle power station is around 35%. More advanced power stations with combined cycles have thermal efficiencies around 45%.

Natural gas-fired CHP stations that recover exhaust gases have overall thermal efficiencies of 80%–90%, and sometimes even higher.

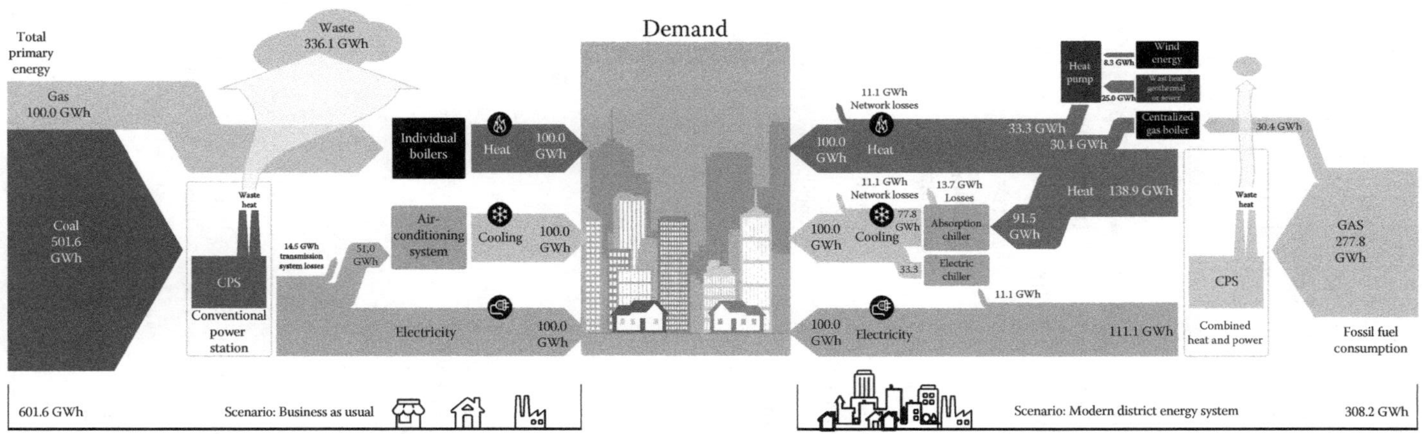

FIGURE 1.2 Economic and environmental benefits of district cooling for a city. (From UNEP, *District Energy in Cities—Unlocking the Potential of Energy Efficiency and Renewable Energy*, United Nations Environmental Programme, 2015.)

This is why the total primary energy utilized in BAU scenarios shown in Figure 1.2 is 601.6 GWh compared to a primary energy utilization of 308.2 GWh with a CHP station. This is a savings of 293.4 GWh or 48.8% compared to BAU, although in both cases, the same energy is produced and taken up by end users: 100 GWh of heat, 100 GWh of cooling, and 100 GWh of electricity.

Table 1.1 explains how these figures are derived with efficiencies stated for each process.

TABLE 1.1
Comparison of Energy Consumption between a Traditional Power Station and a Modern Power Station Using CHP to Produce a Fixed Quantity of Heating, Cooling, and Electric Energy

Case		(1)	(2)	
Power Station		**Traditional Power Station**	**Modern CHP Station**	
Primary energy used, GWh		601.6	308.2	
Source	Coal, GWh	501.6	–	
	Natural Gas, GWh	100	308.2	
Heating energy	Efficiency	99%	Wind energy	8.3
	Coal, GWh	100	Waste heat	25.0
		N/A	By heat pump COP = 1	33.3
		N/A	Recovered heat	47.4
		N/A	Centralized gas boiler	30.4
	GWh	N/A	Total (minus losses[a])	100
Electric energy production other than for cooling	Efficiency	33%	36%	
	By electric energy at customers, GWh	100	By natural gas at customers	100[a]
To air-conditioning systems	At customer, GWh	51.0	At district cooling station	
	Percentage including electric energy other than for cooling	34%	Absorption chillers	COP = 0.85
	COP	2.0		77.8[b]
			Electric chillers	COP = 3.0
				33.3
	At customers, GWh	100	At customers	100
Electric energy, GWh		100[c]	100[c]	

Notes:

[a] Network losses: For case 2 heating: 11.1 GWh, cooling 11.1 GWh.

[b] Absorption chiller losses: 13.7 GWh.

[c] Electric power transmission losses are for cases 1 and 2, respectively: 11% or 14.5 GWh and 11.1 GWh.

High thermal efficiencies were obtained because recovered heat was used to fire absorption chillers and assisted by wind and geothermal heat. District heating and cooling technology is utilized with this modern CHP station.

This is why district cooling and heating is such an important technology. It reduces carbon footprint, increases efficiency of power stations especially when coupled with recovered process heat, and makes use of diversity factors in reducing overall heating and cooling needs.

This is expounded upon in Chapter 3. But district cooling and heating can also be applied at a district level, not only at the power station level. Chapter 2 explains the costs involved in district cooling systems, the tariffs applied, and the economic models used to achieve expected internal rates of returns (IRRs).

Chapters 4 and 5 are concerned with how to design a DC plant and how to distribute chilled water efficiently to end users.

In Chapters 6 and 7, the connections to the end users are explained as well as how to shave peaks off daily cooling loads to reduce electric on-peak energy consumption and make use of off-peak energy storage. Chapter 8 explains the sequence of operations programmed into a DC system, supervisory control and data acquisition (SCADA), a building management system (BMS), and an example for a DC plant philosophy of operation.

The case studies at the end of the book explore district cooling and heating systems, utilizing both the vapor compression chiller technology and absorption technology, and present an example on how to design a turbine inlet cooling (TIC) system for a gas turbine module in a power station to increase its efficiency at peak ambient temperatures in a high-ambient temperature country.

This book is a reference book. It is intended for those studying for an engineering degree and also for district cooling designers/providers, technical people who want to refresh their knowledge about district cooling, practicing refrigeration and air-conditioning engineers, and graduate research engineers in this field or associated fields.

This is not a best practice publication or a guide in this field, rather it is a reference book with practical insights and case studies.

1.3 THE OTHER BENEFITS OF DISTRICT COOLING

Figure 1.2 shows the economic benefits of district cooling and heating at the level of a city power station. District cooling has many other advantages compared to other cooling systems:

1. The net present value (NPV) of a district cooling system, over its lifetime, compares favorably with individual cooling stations for the same district.
2. It utilizes fewer chillers to cool a certain district because of load diversity, especially in mixed-use applications. This saves on capital costs and allows a better base load efficiency.
3. It utilizes thermal energy storage (TES), thus averaging the cooling load of the station, and it helps shave peak electric loads, thus improving electric power consumption.

4. It reduces carbon emissions because of reduced installed capacity and better plant efficiency.
5. It helps free up capital expenditure of property owners and developers to allow them to direct their capital toward their core business.
6. It provides better reliability and dependability compared to in-building individual air-conditioning plants.
7. It has a larger redundant cooling power compared to individual building power stations.
8. It reduces the maintenance cost for individual property owners and developers because the DC provider is responsible for operating and maintaining the DC station(s).
9. It eliminates having to install cooling stations inside or on top of individual buildings, thus improving the aesthetics of the building and freeing-up space that can be used on rentals.
10. It allows for cost savings of electric distributed power cables, conduits, and step-down transformers, when compared with individual building stations, which reflects positively on overall electric installation costs for development.
11. It enhances the possibility of using low energy heat, especially reject heat when available, to produce cooling and heating power.
12. It reduces the liability costs on property owners or developers and associated insurance premiums for providing cooling energy.

1.4 ORIGINS, PRESENT, AND FUTURE STATUS OF DISTRICT COOLING

District heating is not a new idea. It dates back to the Roman Empire when multiple buildings were heated by hypocausts. Earlier artificial incubators were used in ancient Egypt by priests, and the knowledge was used by G. della Porta to make incubators in Naples, Italy, in 1588.

It has been reported that the first recognized commercial district heating system appeared 140 years ago in Lockport, New York, in 1877.[3,4] The new technology spread, and in the next 10 years about 50 new systems were installed.

District cooling was introduced in Hartford, Connecticut, in 1963 by the Connecticut Natural Gas company to utilize natural gas for air-conditioning.[5] District cooling has been available for a much shorter time, about 50 years.

District cooling has benefited from improvements in technologies that have resulted in a renewed interest as an important technology for providing cooling such as

- Improvement in the efficiency of new chillers
- Improvement in the efficiency of distribution systems pumping
- Better prefabricated, preinsulated piping suitable for direct burial, thus cheaper distribution systems
- Increased importance of cogeneration systems that have a higher thermal efficiency (70%–85%)

- Increased importance of trigeneration, the simultaneous production of heating, cooling, and electric energy
- Improvement in the development of large stratified chilled water thermal storage tanks

Asia and the Middle East are showing an increasing interest in district cooling, especially around the Gulf region, where it has grown exponentially since the 1990s. Gulf Cooperation Council (GCC) countries are expecting cooling capacity needs to triple between 2010 and 2030.

Figure 1.3 shows an expected increase from 127 to 352 million kW (36–100 million TR). Figure 1.4 shows potential district cooling expected additions in three GCC countries to be 85.43 million kW (24.27 million TR).

Country	KW (TR), millions
KSA	44.88 (12.75)
UAE	30.20 (8.58)
Qatar	10.35 (2.94)
Total	**85.43 (24.27)**

In Europe, more projects are contemplating district cooling, especially now that an increased awareness of environmental issues has shed light on the benefits of district energy and the recognition that district energy can be an alternative solution to these issues.

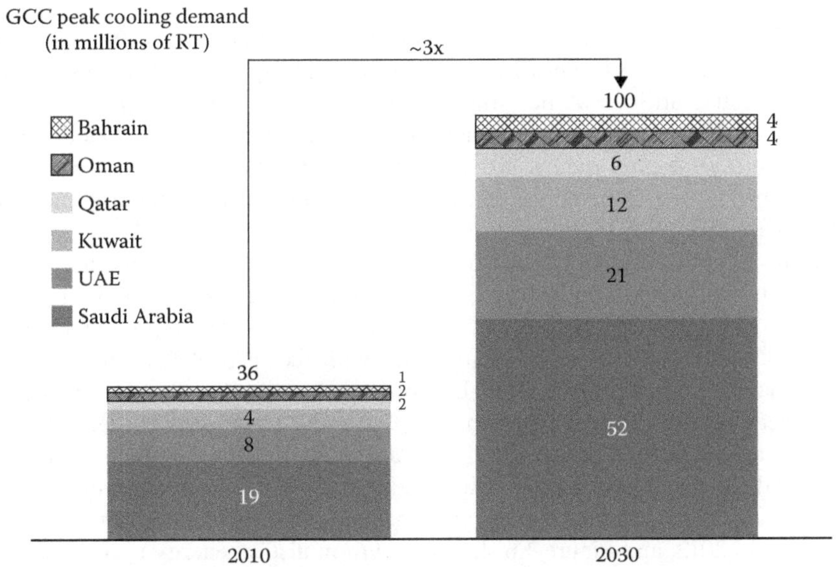

FIGURE 1.3 Peak load cooling demand in GCC countries. (Data from El Sayed, T., et al. *Unlocking the Potential of District Cooling: The Need for GCC Governments to Take Action.* Booz & Co. [now strategy&], New York, 2012.)

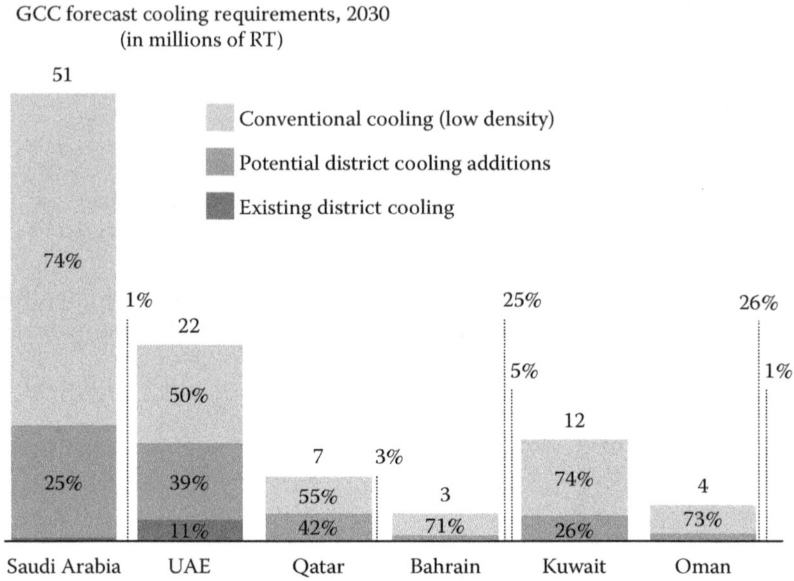

FIGURE 1.4 Forecast potential for district cooling additions in GCC by 2030. (Data from El Sayed, T., et al. *Unlocking the Potential of District Cooling: The Need for GCC Governments to Take Action*. Booz & Co. [now strategy&], New York, 2012.)

The United States is the pioneer of district cooling and heating and has set the pace for many years with innovative technologies. Many U.S. states are deregulating their electric utilities. A building owner is capable of buying electricity from producers other than their local provider. Utilities are creating subsidies that make district cooling systems electric driven, offering customers an alternative service using electric power. Demand for district cooling is expected to increase once these measures are taken.

Canada has recently adopted district heating and cooling systems in some 130 projects.[5] In China and South Korea, there are many large district cooling projects and DC is still a growing industry. India is looking at district cooling and contemplating adopting it in five of its major cities with the help of the United Nations Environmental Programme, UNEP. In Northern Europe, Sweden is a pioneer of district heating and cooling, where both electric and absorption chillers are used. Denmark and Finland operate district cooling and heating systems. Both Great Britain and Germany have used district heating and cooling in many projects.

Figure 1.5 shows district energy growth in North America for new customers from 2008 to 2013; and Figure 1.6 shows growth in district energy in the rest of the world except North America, for new customers during the same period. Schools, hospitals, and health facilities lead the growth in North America while the rest of the world's growth is in commercial and residential sectors.[5]

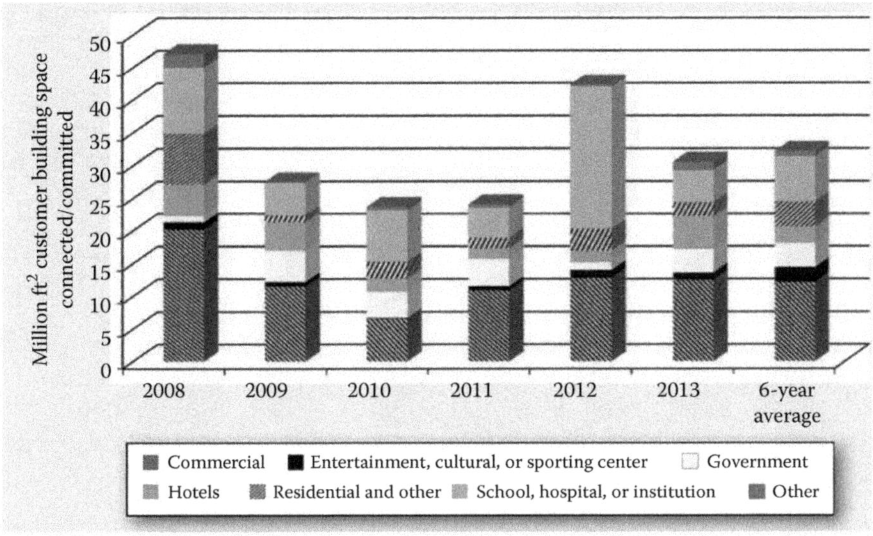

FIGURE 1.5 Growth of district cooling in North America. (From Tredinnick, S., et al. *ASHRAE Journal*, 2015.)

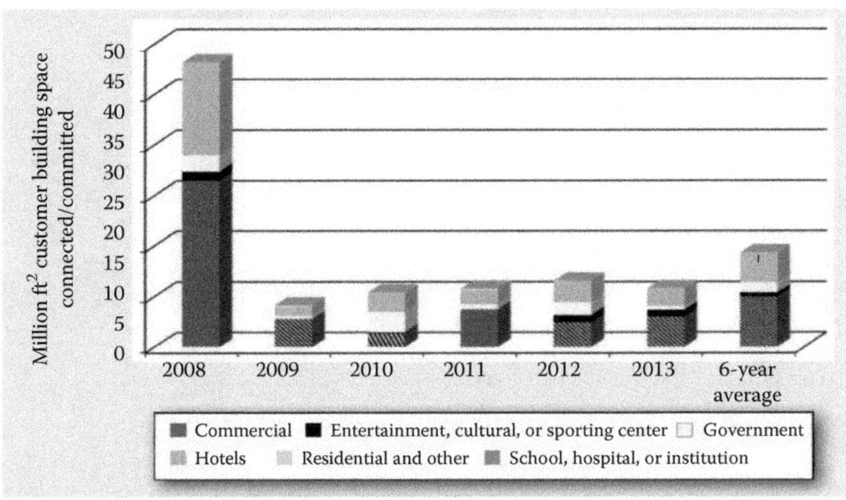

FIGURE 1.6 Growth of district cooling in the world, except for North America. (From Tredinnick, S., et al. *ASHRAE Journal*, 2015.)

District cooling has a promising future and is becoming part of accepted solutions to future challenges facing the world, providing efficient, environmentally accepted energy.

2 Economic Considerations

District cooling schemes are owned by either property developers or a specialized DC provider with the purpose of selling energy to building occupants in exchange for building, owning, operating, and maintaining the system. A specialized agreement between a DC provider and developer is more common with conditions and duration of the contract running for 20–25 years.

The DC scheme comprises the plant and its various equipment, but may or may not include the distribution network, although the property developer is more often than not paying for this part separately. The DC provider will charge the property developer(s) or user(s), depending on the DC contract scheme adopted using usually a three-tariff system rather than one overall tariff system. The tariff structure is explained in this chapter. Tariffs are derived using an economic model with certain assumptions such as depreciation rates, discount rates, debt to equity ratios, and other financial indicators. Their influence will reflect on the internal rates of return (IRRs) of the project.

Afterward, the payback period can be calculated and the corresponding connection, capacity and consumption tariff rates are derived.

This chapter also looks at the need for legislation at a national level to improve the conditions in which DC can flourish and increase the benefits, both economic and environmental, at the country level.

2.1 DC CONTRACT MODELS

There are four typical models for contracts between suppliers and users of a DC system. These are shown in Figure 2.1. The contracting models of DC schemes (shown by Strategy& [formerly Booz & Co.])[1] are explained in Table 2.1.

The type of model used can be with the developer retaining ownership or not, depending on whether constructed units are sold or not. The developer may or may not install submeters to measure energy consumption. In all cases, a DC service agreement is drawn, which guarantees payment to the DC service provider reflecting which party keeps the ownership of the units and whether there are public areas in the development or not.

The four main cases are listed in Figure 2.1 and Table 2.1.

2.2 DISTRICT COOLING TARIFFS

A DC provider will recoup their capital expenditure and operating costs by charging tariffs to the property developer or tenants of the district, or both. These are discussed below.

2.2.1 CONNECTING CHARGE

This tariff is paid only once at the time the contract is signed between the property developer and the DC provider. It is based on the quantity of kW or tons of refrigeration (TR) installed in the connected building.

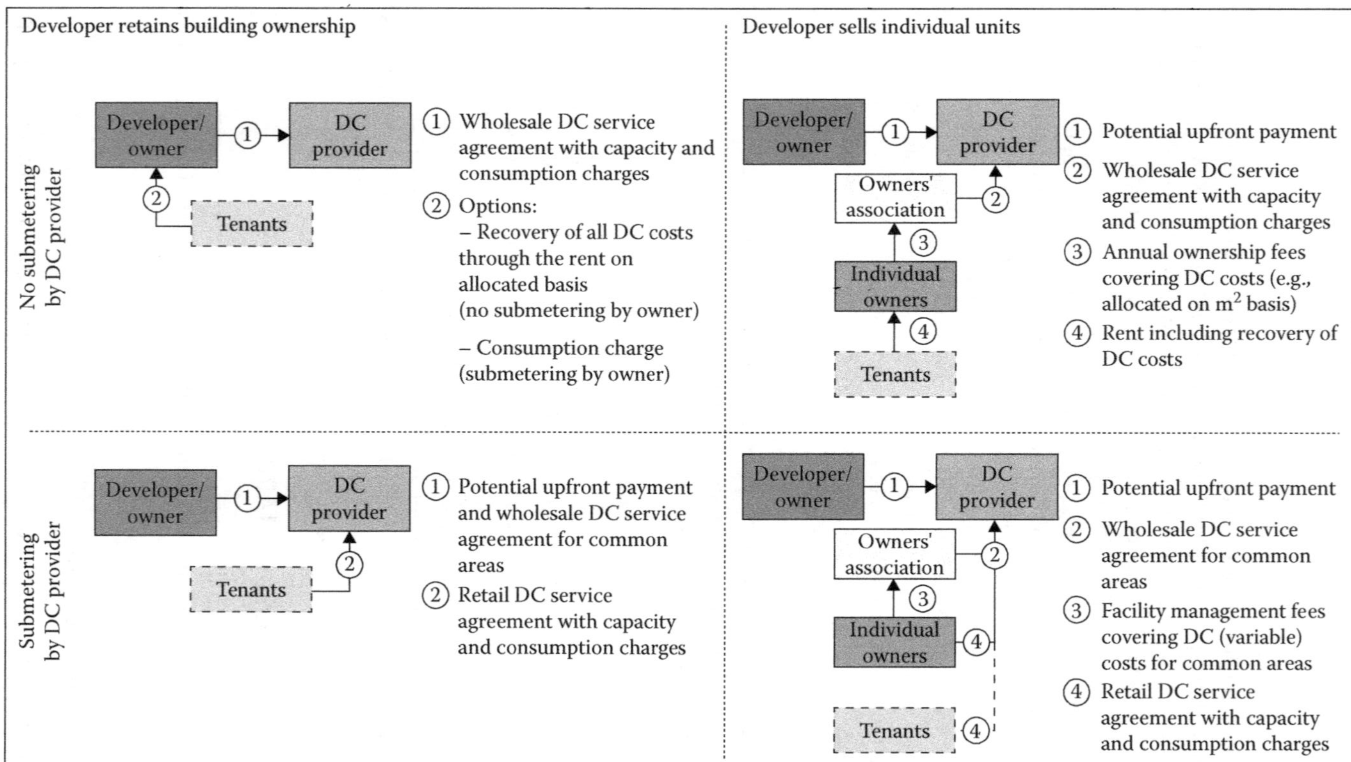

FIGURE 2.1 Contracting models for district cooling schemes. (From El Sayed, T., et al. *Unlocking the Potential of District Cooling: The Need for GCC Governments to Take Action*. Booz & Co. [now Strategy&], New York, 2012.)

TABLE 2.1
DC Contracting Models

Submetering of Consumption	DC Contracting Models	
	Developer Retains Ownership of Units	**Developers Sell Individual Units**
No submeter	1. Wholesale DC service agreement between developer and DC provider with capacity charges. OR 2. Developer recovers all DC costs through rent on allocation basis (no submeter by owner). 3. Developer recovers all DC costs by consumption charges (submetering needed by owner).	1. Upfront payment by developer to DC provider. 2. Tenants pay rent including recovery costs of DC to individual owners. 3. Individual owners pay owner association fees covering DC costs allocated by m^2 basis. 4. Owners' association makes DC service agreement with DC provider with capacity and consumption charges.
With submeter by DC provider	1. Upfront payment and wholesale DC service agreement between developer and DC provider for common areas. 2. For retail areas, a DC service agreement between DC provider and tenants using capacity and consumption charges.	1. Potential upfront payment. 2. Wholesale DC service agreement for common areas between owners' association and DC provider. 3. Facility management fees covering DC costs, variables for common areas between individual owners and owners' association. OR 4. Retail DC service agreement with capacity and consumption charges between tenants and DC provider.

Source: El Sayed, T., et al. *Unlocking the Potential of District Cooling: The Need for GCC Governments to Take Action*. Booz & Co. (now strategy&), New York, 2012.

The tariff pays the cost of connecting the building(s) or tenant units with a connection interface (see Chapter 6), and thus reduces the DC provider's capital expenditure, and therefore the capacity charge. It is somewhat similar to the costs associated with installing an electric meter in a premise.

2.2.2 Capacity Charge

This tariff is paid monthly, starting from the date when the service is ready for use and based on the quantity of kW or TR allocated to the building. The charge is to pay for the capital cost including debt for constructing the DC system over the duration of the contract. This charge is somewhat similar to the demand charge levied by electric utilities.

It is also designed to protect the DC provider in case the property developer or tenant does not use his or her building or unit and subsequently does not consume cooling energy.

2.2.3 Consumption Charge

This tariff is also paid monthly, starting with the consumption of cooling energy and is based on the quantity of energy consumed per kW or TR multiplied by the monthly consumption hours. The charge is calculated by energy meters and is designed to pay for energy consumption, other consumables, and operational manpower. This charge is somewhat similar to the monthly electric bill of the electric utility company.

These two last charges are subjected to yearly increase tied to a yearly inflation rate and yearly index of labor salary increases.

The increases (or decreases) of rates of energy, water, and sewage are calculated using a mathematical formula to be agreed on between the parties before the service contract is signed and are reflected in the tariff rates.

In some cases, the property developer requires a unified tariff for the DC project. In this case, a unified tariff is made based on kW.h or TR.h or per m square air-conditioned floor: This tariff replaces all the three tariffs discussed earlier or may exclude the connecting charge and reflect capacity and consumption tariffs only.

In some countries,[6] additional charges are used to allow better flexibility of usage and guard against low ΔT syndrome and also if the provider is not achieving the chilled water design temperature. Those additional tariffs are charged to the customer or the service provider as follows:

- Charged to the customer
    - *Capacity overrun charge*: Allows the customer to increase his or cooling capacity; calculated as a daily rate equal to one-tenth the monthly rate of the capacity charge
    - *Return temperature adjustment*: Increase of consumption charge by 3% for each degree Celsius of the monthly average return temperature below the system design return temperature
- Charged to the provider
    - *Supply deficiency rebate*: Twice the corresponding capacity charge paid to the customer when the average supply temperature for any hourly interval fails to meet specifications

2.3 PIPING NETWORK LOAD DENSITY, ECONOMIC MODELS, AND IRRs

One of the most economically critical factors deciding the success or failure of a DC scheme is the cost of a piping network and whether the load density in the district is high enough to warrant the construction of a piping network.

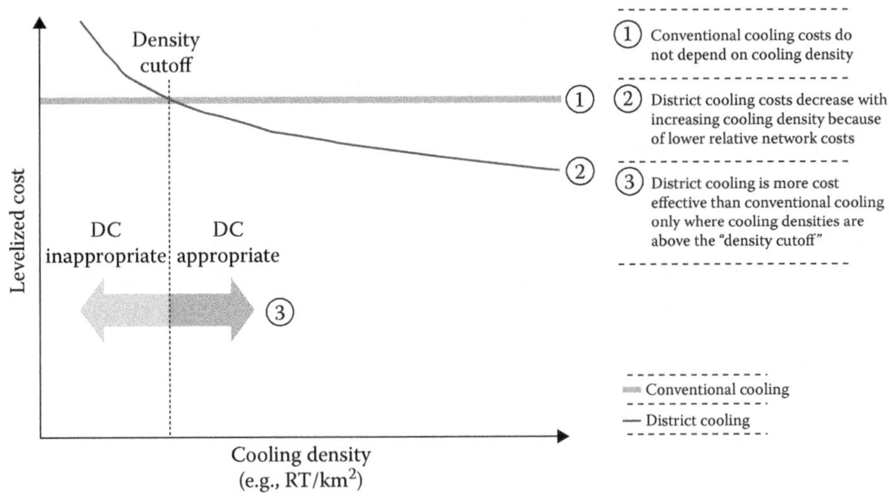

Note: RT/km^2 = Refrigeration tons per square kilometer. Levelized cost = Price required to breakeven.

FIGURE 2.2 Levelized cost versus cooling density for district cooling schemes. (From El Sayed, T., et al. *Unlocking the Potential of District Cooling: The Need for GCC Governments to Take Action*. Booz & Co. [now Strategy&], New York, 2012.)

Figure 2.2 shows the levelized cost (price required to breakeven) of a district cooling system versus cooling density (in RT per km^2). For a district cooling scheme to be successful, those levelized costs must be less than the costs of a conventional distributed in-building air-conditioning station. The less those costs are the better the economic return of the system, as the figure shows. Those costs usually occur at a certain cooling density over 35,000 kW (10,000 TR) per km^2 for a mixed use development in a high-ambient temperature country.

Another important factor that can affect the success of a DC scheme is the number of Equivalent Full Load Hours (EFLH) the system performs. This is explained in detail in Chapter 3. This figure shows that the EFLH is dependent on the shape of the daily cooling load demand curve and how often large loads occur during a typical year.

In short, once EFLH is established for a system, aggregating the various activities of the building's daily cooling load demand curves will make it possible to find the average daily load and choose the type and size of a thermal energy storage (TES) tank to use. The average daily load will also make it possible to choose the total capacity of chillers needed, their numbers, and each chiller capacity. The prices of energy and their availability will help decide the type of chillers to use, whether they are mechanical vapor compression chillers, absorption chillers direct-fired, hot water, or steam-fired or by recovered exhaust heat chillers, or a combination of all three (see the Case Study: TIC for a Power Station).

Once this is established, the size of the machine plant room can be calculated, ancillary equipment selected (such as cooling towers and pumps), and a reasonably accurate cost established for the plant and distribution network (if it is within the scope of the DC provider). With capital costs established and other parameters also established, operating costs can be estimated and major parameters of a business model are set.

The length of a DC provider contract is usually around 20–25 years, although this may change depending on the agreement between the parties.

To evaluate the tariffs with which a DC provider can recover his or her Capex and Opex costs, a net present value (NPV) computer program is used that encompasses all costs and charges with an economic comparison between in-building air-conditioning stations and the district cooling scheme for the full duration of the contract.

ASHRAE Applications Handbook[7] (2011, chapter 37 "Owning and Operating Costs") as well as ASHRAE HVAC Systems & Equipment Handbook[8] (2012, chapter 12 "District Heating and Cooling") provide detailed explanations of the calculation of life-cycle cost analyses.

A certain IRR is chosen for the model. This depends on the rate of borrowing, the debt-to-equity ratio, discount rates, insurance rates, and other important factors. Connecting, capacity, and consumption charges are then derived and adjusted to achieve those IRRs and achieve a NPV for the system that shows financial savings using DC as opposed to using individual in-building stations.

In this way, it can be shown that the district cooling scheme will provide a better proposition economically to the end user than individual in-building stations as well all other benefits described in Chapter 1.

2.4 ENERGY SUBSIDIES AND THE NEED FOR LEGISLATION AT A NATIONAL LEVEL

Figure 2.3 shows two graphs of the cooling costs of a system versus cooling density using actual energy costs and energy costs with a government subsidy.[1] The graphs show that energy cost subsidies will shift the actual cutoff point to the right, creating

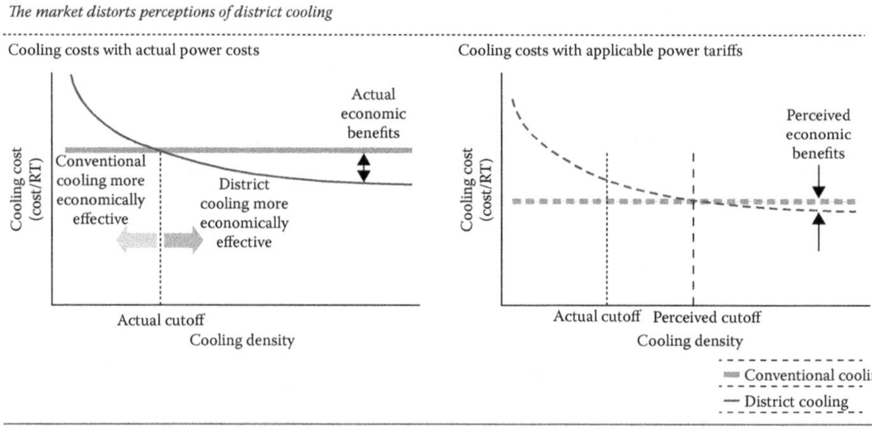

FIGURE 2.3 District cooling costs and conventional cooling with and without energy subsidies versus cooling density. (From El Sayed, T., et al. *Unlocking the Potential of District Cooling: The Need for GCC Governments to Take Action*. Booz & Co. [now Strategy&], New York, 2012.)

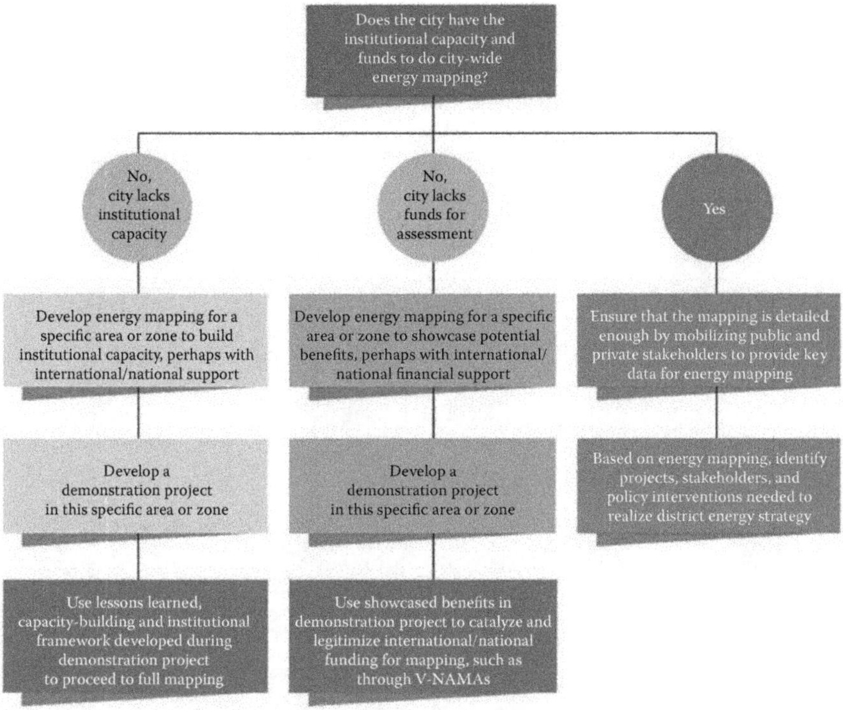

FIGURE 2.4 Energy mapping of a city. (From UNEP, *District Energy in Cities—Unlocking the Potential of Energy Efficiency and Renewable Energy*, United Nations Environmental Programme, 2015.)

a larger cooling density compared to that at actual prices of energy. This will also diminish the economic benefits of a DC scheme.

In order to address this point and give DC the ability to maximize the economic benefits for a country, the government must address those issues by reducing subsidies and move to provide direct cash subsidies to deserving individuals, so that it does not disrupt laws of supply and demand and damage the country's overall energy policy.

Figure 2.4 shows the steps needed to map energy in a city when institutional capacity and funds are lacking. This is the first step to developing an energy strategy and hence identifying targets to obtain a complete heating and cooling assessment.

Demonstration projects would need to be developed for a certain area or district, and the institutional framework that needs to be developed to obtain a full energy mapping should be learned by this process.

Figure 2.5 shows the steps at local and national levels needed to develop a district energy policy utilizing available city resources of planning, regulating, facilitating, conditioning, advocating, consuming, and providing district energy (DE). At the local level, taxes are proposed and incentives are provided, adding DE in energy

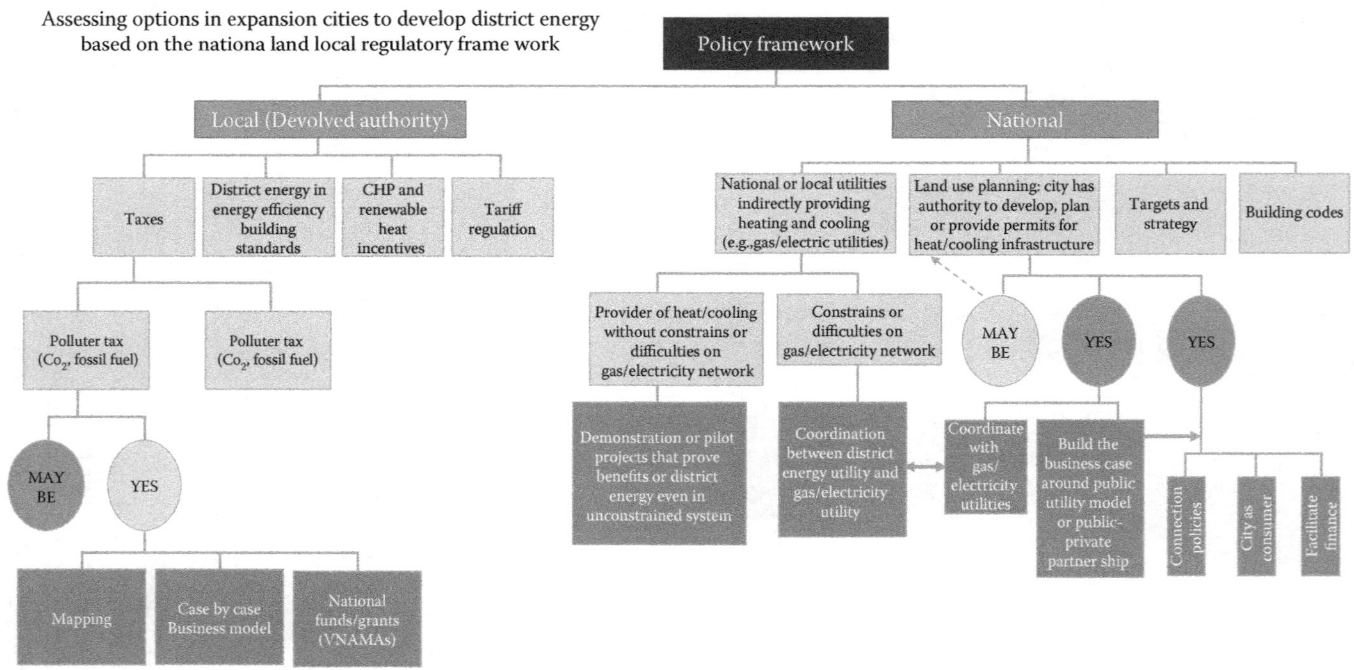

FIGURE 2.5 Policy framework at local and national levels to develop district energy in a country. (From UNEP, *District Energy in Cities—Unlocking the Potential of Energy Efficiency and Renewable Energy*, United Nations Environmental Programme, 2015.)

efficiency building standards and tariffs regulations. On the national scale, national or local authorities may choose to provide heating and cooling services through national or local utilities.

Also on a national scale, city land-use planning is developed to authorize developments and issue permits for heating and cooling infrastructures. The adoption of a district cooling and heating strategy also requires the adoption of building codes, including a district cooling/heating code. These measures together with attainable targets will ensure achieving the goals of an energy strategy.

3 Major Factors Influencing the Design of a District Cooling System

3.1 DEFINING THE DISTRICT AND PLANNING

3.1.1 DEFINING THE DISTRICT'S BOUNDARIES

Defining the boundaries of a district cooling system at the beginning where the provision of chilled water is intended, is of paramount importance. Failure to define the district boundaries can create problems in the future, such as an undersized or oversized refrigeration capacity of the plant. This also applies to sizing and routing the chilled water-piping network. Failure to properly define the boundaries may also lead to conflicts between the district cooling provider and the users/developers, resulting in disruption of service and financial losses to all parties involved.

Even the best planned district cooling project may suffer from delayed construction of some of its buildings. When a clear definition of boundaries is made, the resultant problems associated with delayed construction and loss of income can be known early on, and therefore expected operational losses can be dealt with and factored in the plant and network construction.

Obtaining local government permits for construction and municipal agency permits for electric power, fresh water, and discharge water is time consuming. Therefore, the sudden demand of unplanned added refrigeration capacity due to poor definition of the district boundaries at start may necessitate reapplying for permits and may result in loss of time and expenses. The integration of the chilled water-piping network with other utilities, such as water supply, sewage, electric supply, and natural gas supply, can also cause disruption because of poorly defined district boundaries.

3.1.2 THE MASTER PLAN, BUILDING LOCATION, AND DEVELOPMENT RATE

In order to estimate the cooling load of a district, it is important to examine carefully the master plan at an early stage. The district cooling provider should work early on with the developer to identify the size of the buildings, their function, cooling load estimates, and other possible special features. This allows the building's cooling load requirements, their peak load timing, and duration to be included in the overall cooling load estimate of the district.

It must be understood that a district cooling master plan will not always reflect the final development shape. The developer will often accommodate changes in the master plan to include late requirements of clients after the piping network is routed, designed, and executed. The district cooling provider will deal with those changes by trying to make some changes in the chilled water piping, if possible, without drastically changing the actual design, or provide those developments with a temporary chiller plant (a modular central plant) until permanent chilled water piping can be constructed.

The development rate of a district is usually defined clearly prior to starting the design of the system. This rate is conditional on the number of buildings, their type, and the date when those buildings are expected to be ready to receive the service. Often there are delays in construction, and the buildings are not ready to receive the service at the defined dates. This causes financial losses to district cooling providers. The opposite is also true when a building in a certain location requires chilled water at an earlier date than planned and the chilled water-piping network is not yet constructed in this area. It is therefore important for the district cooling provider to update its master plan regularly and try to anticipate the impact of changes on the installed capacity of the plant and the routing and design of the chilled water-piping network.

3.1.3 Quantity and Location of DC Plants in the Master Plan: Topography, Load Demand Density Distribution, and Piping Route

The initial master plan of a development is the basis for designing a chilled water-piping network. The location of each building, its estimated cooling load requirement, and its purpose will define a daily cooling load demand profile.

Deciding on the location of the DC plant requires working closely with the developer. The topographic nature of the land—the elevations and depressions, the soil composition, and other geographical aspects—will influence location of the plant(s).

Locating the district cooling plant near the heaviest cooling loads is advantageous to the district cooling provider because of savings in piping tubing and sizes, excavation and construction costs, and eventually operating pumping expenses.

Developers may object to those locations or its interference with the initial master plan. Plants are not usually located where a district cooling system can be served best, but in isolated periphery areas, which results in extra expenses in the initial piping cost and operating expenses.

3.1.4 Permits and Interference with Other Utilities

In many high-ambient temperature countries, especially those that are developing at a high rate of growth, the advantages of district cooling systems are not well understood. Although district cooling is a cornerstone for curbing increasing electrical power demands in those fast developing countries and for carbon dioxide emission reductions (see Figure 1.2), little has been done to create mechanisms to help this industry.

Unless government departments responsible for issuing permits for district cooling projects have guidelines regulating utilities planning, district cooling companies will struggle to obtain and execute their projects. The problem becomes even more compounded because district cooling projects are usually built in cities and towns that are densely populated, which defies the conventional planning logic of not allowing mechanical plants in those areas. There is a need to create local mechanisms that explain the benefits of district cooling to central and local agencies. The early adoption of guidelines by those government agencies for utilities planning and integration will facilitate permitting and construction of district cooling systems (see Figures 2.4 and 2.5).

3.2 DIVERSITY FACTORS AND THEIR DECREASE AS THE DISTRICT DEVELOPS

Individual buildings peak at different times. This is why the coincident overall peak demand of a district cooling system is not the sum of each individual building peak demand.

Diversity factors are used to calculate the overall peak load of a district cooling system. Those diversity factors may be as low as 0.6 or 0.7 of the sum of individual building peak demands in applications where there is a great diversity of use.

There are different types of diversity factors. Diversity factors inside a building depend on the actual use pattern of a building. Diversity factors between one building and the other in a district depend on each building's function, orientation, use, and diversity factors between district cooling plants that may be serving a single district's distribution network.

Chilled water-piping networks are also subject to diversity factors between distribution loops serving different buildings in parallel. All those diversity factors must be taken into account when calculating the overall peak demand of a district cooling system and when designing chilled water distribution networks.

Diversity factors are often calculated using the building's daily peak load demand profiles. Otherwise, established rule of thumb figures are used. It is important to establish diversity factors prior to calculating the district cooling system's overall installed refrigeration capacity in order not to overestimate this capacity. Overestimating a plant capacity will result in financial losses because of unutilized installed capacity and adversely affect the expected revenues.

Underestimating a plant's refrigeration capacity causes insufficient cooling load availability and the loss of redundant capacity. This may lead to an inability to meet customers' cooling load requirements and result in financial losses to both the user and operator.

At its start, a district cooling system rarely operates at its maximum intended capacity. It is usual for a system to start with a much reduced capacity and install more capacity as more and more building construction is completed and chilled water demand increases. During this "maturity" period, the diversity factors are high. As buildings are completed and occupied, the type of building function will be more pronounced and may vary. Diversity factors start to decrease and reach their lower limits.

3.3 THE DAILY COOLING LOAD DEMAND CURVE AND PEAK LOADS

Several important factors must be clearly defined when designing a district cooling system. Some of the most important factors are the daily cooling load demand curve and peak loads. A customer design engineer or consultant usually defines a building's cooling load.

These buildings could be administrative, shopping malls, hotels, schools, and other types of buildings. Cooling load estimates of these buildings will usually vary a great deal from building to building. An administrative building's cooling load estimate will probably include loads attributed to the prevalent weather, loads of occupants, electrical and electronic appliances, lighting, and other loads. These cooling load estimates will differ from those of a shopping mall where the occupant's load will probably constitute the major part. The same applies to other buildings as well where the loads will vary a great deal. Shopping mall loads peak at a different time of the day compared to administrative loads or residential loads.

Deciding how large and when those loads occur is of crucial importance in calculating the total design load of a district cooling plant.

In estimating the cooling load of buildings for a certain district, it is possible to use computerized simulation programs and thus obtain an accurate understanding of peak loads' occurrence and their magnitude.

Figure 3.1 shows typical daily cooling demand profiles for buildings in a high-ambient temperature country. Peaks differ according to the application. Recreational applications tend to have peaks late in the evening, whereas occupation-related applications peak in the afternoon. The weekend operational hours are usually much reduced and different from those during the weekdays. An aggregate daily cooling load demand curve can thus be drawn, which shows the overall daily system load demand curve. Figure 3.2 shows a typical curve. This is an important curve profile as it defines the peak refrigeration capacity of the district cooling plant, when the peak occurs, and how long it lasts decides the nature of the cooling load.

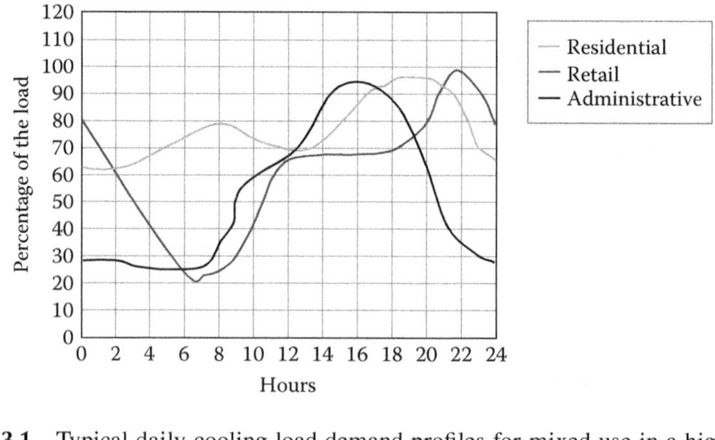

FIGURE 3.1 Typical daily cooling load demand profiles for mixed use in a high-ambient temperature country.

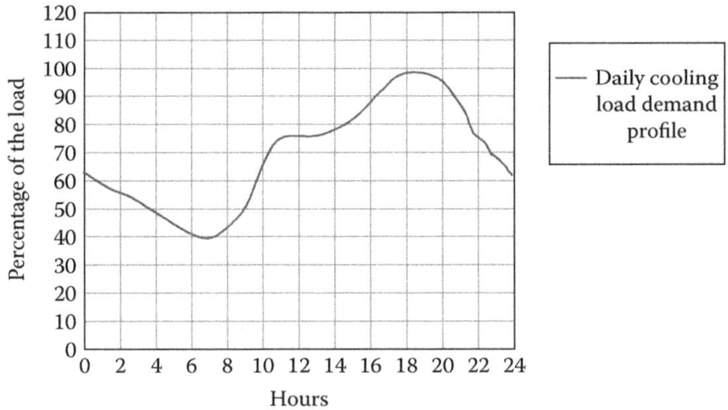

FIGURE 3.2 Typical aggregated daily cooling load demand profile for mixed use in a high-ambient temperature country.

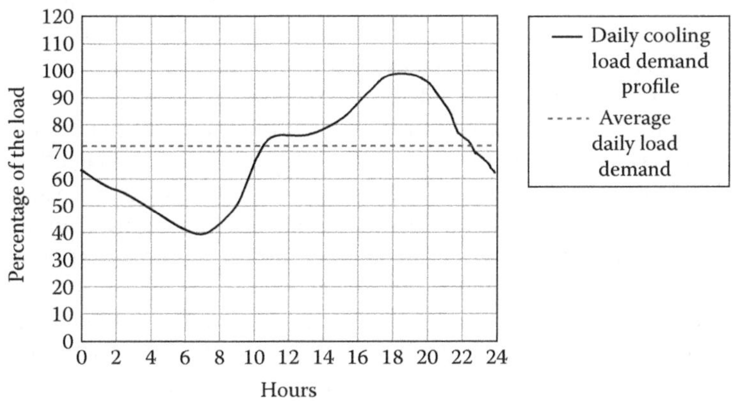

FIGURE 3.3 Typical daily cooling load demand curve for mixed use, showing the average daily load.

Figure 3.3 illustrates a typical daily cooling demand curve for a mixed use district cooling system. Note the horizontal line that shows the average daily cooling load. The thermal energy storage (TES) will have to be designed to suffice the load above this line during discharging and charging over the period below the line. These values will also help choose the size of the TES tank and define its charging or discharging cycle time (see Chapter 7).

Figure 3.4 shows typical daily cooling demand profiles for several superimposed applications: office, residential, hotel or service apartments, shopping, and leisure (by the kind permission of www.araner.com). The curves show the difference in demands according to applications and therefore the potential diversity and thermal storage.

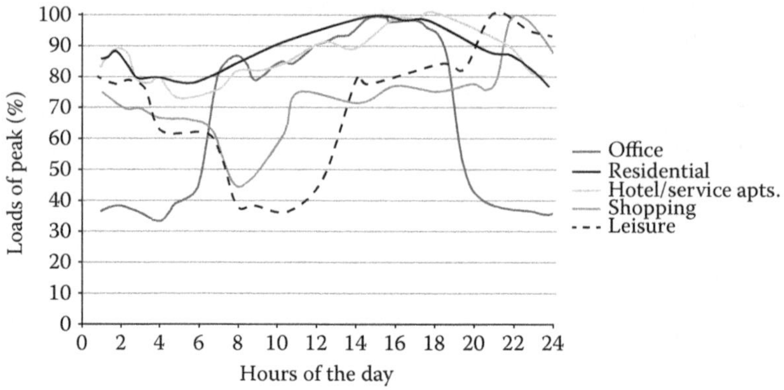

FIGURE 3.4 Typical daily cooling load demand profiles for several applications, superimposed. (Reprinted by permission from www.araner.com.)

Note the following trends for each application:

- *Offices*: There are two peaks, one in the early morning (6–10 a.m., about 85%) and a larger one in late afternoon (2–6 p.m., about 95%). One of the well-defined profile characteristics is as follows: peaks occurring between 6 and 8 p.m. with the basic load around 38% the rest of the day.
- *Residential and hotel/service apartments*: Profiles are somewhat similar in general trends. There are two peaks, first peak (1–3 a.m., about 90%) and a second peak (1–8 p.m., about 100%), reflecting the late habits, because of the severe heat during the day for hot ambient temperature country residents. For more moderate temperature countries, the first peak would be expected to shift to the right by about 4 h.
- *Shopping and leisure*: Profiles are somewhat similar, although leisure application shifts are more pronounced. Two peaks are shown, first one (10 a.m. to 2 p.m., about 70%–80%) and a second peak (8–11 p.m., about 100% plus). Two peaks reflect social behavior related to lunchtime activity and late shopping and recreational activities after working hours.

These different load demand profiles clearly show the diverse nature of the overall aggregated load profile, underlining the importance of load diversity in creating a smaller installed capacity, thus saving in the capital cost.

It also shows the importance of TES when applied to district cooling projects with large diversity, thus saving on operating costs (TES charging at night when temperatures are milder) and on capital costs (TES usually costs less than the same installed capacity for a DC station).

3.4 THE ANNUAL COOLING LOAD DEMAND AND THE EQUIVALENT FULL LOAD HOURS

In estimating a district cooling capacity, the design engineer would often overestimate this capacity. This may be acceptable when designing for an individual building to guard against the harm of an insufficient capacity, loss of redundancy, and inability to reach design conditions at certain peak load times, but for a district cooling system, it can cause a great deal of financial losses to the provider and user of the system. From the point of view of the provider, the inflated calculated revenue stream will overestimate the break-even point of return on investment and thus create losses. The inflated initial capital cost for the plant and chilled water system, will also create financial losses as it will take longer than predicted to recoup cost, on top of reduced operating efficiency of the plant, which increases operating costs.

The user is also paying for this mistake because it will create eventual disagreements between the user and provider.

It is therefore important to find out the annual cooling load demand and the annual Equivalent Full Load Hours (EFLH) for a district cooling system as accurately as possible, thus guarding against those problems.

The daily cooling load demand curves, when put together for a year, will provide an annual load–duration curve.

Figure 3.5 shows a typical annual load–duration curve for a high-ambient temperature country. This curve allows the calculation of the annual refrigeration energy of a system and thus provides, together with the peak capacity of the system, the EFLH of system.

$$\text{EFLH} = \frac{\text{Annual refrigeration energy consumed (ton h)}}{\text{Peak capacity of the system (ton)}}$$

There are historical EFLH figures for many high-ambient temperature countries. Those vary from 1800 to 4700 h, depending on the nature of use of the system and the prevailing ambient conditions.

EFLH figures are important when constructing the business model of a district cooling system, because, together with the installed capacity, they provide the revenue stream of the district cooling system. EFLH figures help decide the most economical type or a combination of types for the cooling system, according to the cost of its operation.

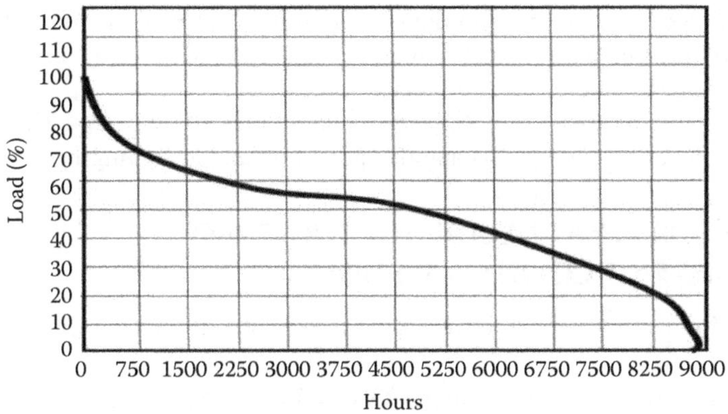

FIGURE 3.5 Typical annual load—duration curve for mixed use in a high-ambient temperature country.

3.5 CHOOSING A ΔT FOR THE SYSTEM AND LOW ΔT SYNDROME

ΔT is the difference between chilled water supply and return temperature. It usually refers to the difference in temperature when measured across the plant supply and return piping. Designers strive to achieve ΔTs as high as possible to reduce capital costs of distribution systems. Increasing ΔT will reduce flow rates of a chilled water system. This in turn reduces pumping energy costs and improves operating costs.

3.5.1 Low ΔT Syndrome

A common problem of district cooling systems is their inability to adhere to ΔT design conditions. This phenomenon is called "low ΔT syndrome." Low ΔT syndrome is wasteful because it causes the system to pump excess rates of flow to satisfy building demands, although the plant is unable to operate to its full installed capacity. This may eventually lead to deteriorating comfort conditions inside the buildings. There are limitations on the safe use of low chilled water temperatures. Mechanical vapor compression systems can operate safely down to about 3°C chilled water supply temperature, without damage to its equipment, whereas absorption refrigeration chiller can safely operate down to about 5.5°C safely (some manufacturers of new-generation absorption chillers claim to operate at lower temperatures, although those new absorption chillers are not yet commercially available on a wide scale). The high return water temperatures also have their limitations. Those should not exceed the dew point temperatures necessary to achieve design conditions in buildings, otherwise dehumidification at the air-side coils suffers and the latent heat capacity of the building's fan coils and air handling units are affected.

A high ΔT design is generally economical to the operation of a district cooling station, the chilled water distribution network, and individual buildings' heating, ventilating and air conditioning (HVAC) systems. This is because of savings in the size of piping and accessories in the plant and larger savings in piping, preinsulation, and accessories in the chilled water distribution network. Savings are also achieved inside the mechanical rooms of building or building groups in energy transfer stations (ETSs) (see Chapter 6). The HVAC system inside the building also becomes economical because of savings in the size of chilled water piping, insulation, accessories, and ducting network, although chilled water coils may become costlier because of added rows to cope with larger ΔTs.

Typical ΔTs for a district cooling system are about 9–12°C (16–22°F). Those are ΔTs that achieve an economical capital expenditure and low pumping operating cost.

3.6 FACTORS THAT DECIDE THE SUCCESS OF A DISTRICT COOLING SYSTEM

The following factors affect the success or failure of a district cooling system.

3.6.1 Initial Cost of the Chilled Water-Piping Network

The initial cost of the chilled water-piping network and its operating expenses may be a very influential factor in deciding the success or failure of a system. The initial cost of the piping network may constitute 40%–60% of the overall cost of the district cooling system; balancing the cooling load positions on the chilled water network is important for reducing the pumping cost. Careful planning is required when designing the chilled water-piping route.

3.6.2 Selection of Chilled Water ΔT

The selection of the chilled water ΔT can affect greatly the initial cost of the piping network. ΔT also affects the pumping cost of the chilled water system.

3.6.3 Calculating the Plant Base-Load and On-Time Additions

The accurate calculations of the plant base-load and on-time addition of new refrigeration capacities help make the system economically successful.

3.6.4 The Accurate Prediction of Diversity Factors

The effective calculation of an integrated diversity factor for the system guards against the often encountered oversizing of the refrigeration capacity of the plant.

3.6.5 Adding an Energy Storage System

The use of TES to provide peak load refrigeration capacity is important to help reduce the initial cost and running cost of the system.

3.6.6 The Load Density of the System

The load density in a district is usually a deciding factor in the economic feasibility of a district cooling system. Residential applications are often not economical. This factor was addressed in more detail in Chapter 2.

3.6.7 The District Cooling Plant's Interface with Building Systems

When a DC plant is capable of instantaneously reading pressure, temperatures, and chilled water flow in a building's mechanical rooms, it can adapt to the chilled water supply and flow much more effectively, compared to the plant operator relying only on distribution network pressure and temperature readings in selected points. The system is then more efficient because it can respond to changes in flow and temperatures more effectively, resulting in savings in operating costs and reduced tariffs.

Many DC schemes suffer from low ΔT syndromes resulting in an increased consumption of pumping energy and inability to achieve chiller design loads. When the operator of a DC system is aware of supply and return chilled water temperatures and flows at each building limit, he or she can tackle low ΔT syndromes much more effectively and save valuable pumping energy.

Chapter 5 explores the various distribution network-pumping schemes. One of the most attractive pumping schemes is the primary-distributed secondary system, where dedicated secondary network pumps are discarded and a secondary pump is located at each building's mechanical room. When this system is applied and all its aspects are well understood, it consumes less pumping energy overall than other systems, and operating cost savings are achieved. If the DC operator can also control as well as monitor those secondary pumps at building's mechanical rooms, the system will operate much more effectively. End users may object to the DC operator authority over their building's pump and interference with building's operation, albeit in a minimal way. This may be the case in systems without plate heat exchanger connection and therefore without total isolation between the network's chilled water and the building's chilled water. If these issues are settled at the beginning, the primary-distributed secondary pumping arrangement is the most suitable arrangement for larger district cooling systems.

4 Designing Central Plants

4.1 TYPICAL DISTRICT COOLING PLANT COMPONENTS

Chilled water is produced at one or more central plants. Central plants comprise primarily the following components:

- Chillers
- Pumps
- Thermal energy storage (TES) tanks
- Cooling towers
- Valves
- Control equipment
- Expansion tanks and air separators
- Water filtration systems
- Air separators and eliminators
- Chemical treatment systems
- Piping and insulation

Designing a central plant includes the proper selection of components in order to ensure that all parts work together in harmony to achieve design goals. These goals are as follows:

- Achieving design efficiencies to reach energy consumption goals
- Achieving water consumption goals, especially in high-ambient temperature (HAT) environments
- Achieving design goals of useful life expectancy of equipment
- Achieving ease of operation and maintenance

In order to achieve these goals, designers should examine the performance of various components at design conditions and ensure that they can reach those stated goals when operating together.

It is important to emphasize the importance of these basic goals. The best central plant layout will not deliver expected design conditions if selected chilled water or cooling water pumps cannot achieve flow rates at stated efficiency goals. In addition, chillers that operate at efficiencies lower than design efficiencies, because of faulty selection, waste energy. Similarly, cooling towers that cannot achieve condenser cooling temperature design values because of poor selection will not achieve chilled water temperatures when needed most. Although the importance of choosing a suitable ΔT for the system is discussed in Section 3.5, it is worthwhile emphasizing that the designer should strive to achieve ΔTs across the plant supply and return chilled water to reduce the cost of the plant piping network, accessories, and insulation. Also to a larger degree, reduce the cost of distribution network, the size of energy transfer stations (ETSs), and the size of heating, ventilating and air-conditioning (HVAC) equipment inside buildings.

4.2 THE IMPACT OF THE MONTRÉAL PROTOCOL ON SELECTING REFRIGERANTS FOR CHILLERS

4.2.1 The Montréal Protocol and Ozone Depletion Substances

In the early 1970s, three scientists, Paul Crutzen, Mario Molina, and Sherwood Rowland, found a connection between the breaking apart of chlorofluorocarbons (CFCs) in the stratosphere and the destruction of the ozone layer. The hypothesis was that CFCs when subjected to UV radiation could produce chlorine radicals that destroy a large number of ozone molecules and deplete the ozone layer. This is the layer in the upper atmosphere of Earth responsible for the protection of life on Earth from harmful UV rays, and its thinning allows UV rays to penetrate to Earth's land masses and oceans, thus increasing risks of skin cancer and eye damage in humans and affecting both the ocean's life and fish's life.

The hypothesis of Rowland, Molina, and Crutzen was ignored until in 1985 an "ozone hole" was discovered over Antarctica. Action was immediate and in the same year the Vienna Convention was held, and in 1987 the Montréal Protocol (MP) was agreed on to eliminate ozone depletion substances.

Because CFCs were, at that time, the most widely used substances as refrigerants for chillers, aerosol propulsion agents, foam-blowing agents, industrial and commercial solvents, sterilants, and other applications, their impact on ozone depletion was great. Hydrochlorofluorocarbons (HCFCs) were used to replace CFCs, especially as refrigerants in refrigeration, air-conditioning, and foam-blowing applications. Other substances depleting the ozone layer were added to MP, such as carbon tetrachloride, a solvent used in the chemical and electronic industries; methyl chloroform, also a solvent; halons and hydrobromofluorocarbons (HBFCs), firefighting agents; and methyl bromide, a fumigation pesticide used in agriculture.

Although MP was originally conceived to phase out CFCs, its subsequent amendments enlarged its mandate and included in total 96 ozone-depleting substances (ODS).

To establish a benchmark to measure ozone depletion, ozone-depleting potential (ODP) was created, and CFC-11 and CFC-12 were given a reference ODP of 1. Other substances were assigned an ODP weight compared to these two in how much more or less they would deplete the ozone layer. Methyl chloroform, for example, has an ODP of 0.1; this means that 10 tons of it would have the same impact as 1 ton of CFC-11 or CFC-12. In the second meeting of the parties to MP (London, 1990), a criterion was created to differentiate between developed and developing countries (Article 2 and Article 5 countries, respectively) in their obligations toward MP. Any country whose calculated annual level of consumption of controlled substances in Annex A of MP is less than 0.3 kg per capita per year, on the date of the entry into force of the protocol, or anytime thereafter until January 1999, shall be entitled to delay for 10 years its compliance with the control measures of MP.

The developed countries (Article 2 countries) had agreed to schedules for the gradual phasing out of ODS in their own economies. They agreed to contribute to a fund that would pay for the costs of transferring new, ozone-friendly technologies to Article 5 countries. In addition, developed countries also agreed to contribute technical assistance as well as share best practices and implementation expertise to ensure that funding assistance was applied effectively.

The Multilateral Fund (MLF) was established on January 1, 1992, with its secretariat located in Montréal, Canada. The MLF's objective is to provide financial assistance to projects that would help Article 5 countries comply with their obligations under the protocol to phase out ODS in use.

The work that the MLF finances in recipient countries is implemented by four implementing agencies: the United Nations Industrial Development Organization (UNIDO), United Nation Environmental Programme (UNEP), and the World Bank. The MLF received contributions of approximately U.S.$ 2 billion, between 1991 and 2005, from 49 industrialized countries. National Ozone Units (NOUs) have been established in 131 countries as government focal points for implementation of this environmental agreement. MP has been a global success, and a massive reduction in ODS use worldwide was achieved. By September 2007, 97% of ODS were phased out. By 2010, CFCs were phased out completely, and MP became the most successful environment agreement of all.

4.2.2 The Montréal Protocol and Climate Change Mitigation

HCFCs were used to replace CFCs. The most dominant HCFC for refrigeration and air-conditioning applications is R-22. It became important to also know the global warming potential (GWP) of the refrigerants used as an alternative to CFCs.

The GWP of a gas is how much heat or greenhouse gas it traps in the atmosphere relative to a similar mass of carbon dioxide (CO_2). The higher the GWP of a gas, the higher it contributes to Earth's global warming, thus changing Earth's climate. Changing Earth's climate means melting the polar caps, a rise in the sea level that leads to flooding of coastal areas, coral reef bleaching, and other phenomena that affect greatly our planet and human life on it. Table 4.1 shows the GWP and ODP

TABLE 4.1
GWP and ODB of Several Refrigerants

Substance	Global Warming Potential (GWP, 100 Year)	Ozone Depleting Potential (ODP)
CO_2	1	0
CFC-11	4.750	1
CFC-12	10.900	1
HCFC-22	1.810	0.055
HCFC-141_B	725	0.11
HCFC-142_B	2.310	0.065

of several refrigerants that were commonly used as alternatives to CFC-11 and CFC-12.[9] The table shows that HCFC-22 has a much smaller ODP than CFC-11 and CFC-22 (0.055 compared to 1), yet its GWP is 1810 times that of CO_2 (1810 compared to 1). Because of the dramatic increase in the production and consumption of HCFCs to replace CFCs over the past two decades, the threat to the ozone layer and climate change has become imminent.

A schedule for phasing out HCFCs was agreed between the parties in 2007 (Decision XIX/6) to accelerate the phase-out schedule for HCFCs.

This ensures faster protection to the ozone layer and also faster mitigation of CO_2e (carbon dioxide equivalent), assuming that alternatives can be found with a low or no climate impact.

The Technology and Economic Assessment Panel (TEAP) of the UNEP estimated in 2008 that 4–5 billion tons of CO_2 is in ODS banks in developing countries and 12–13 billion tons of CO_2 is in ODS banks in developed countries. Banks are the total amount of substances contained in existing equipment, chemical stockpiles, foams, and other products that are not yet released to the atmosphere.[10] Those not yet released to the atmosphere. Those banks may leak to the atmosphere and threaten the global climate. The same decision XIX/6 instructed the MLF to finance pilot projects in developing countries for the destruction of ODS banks. The accelerated phase-out schedule of HCFCs for non-Article 5 and Article 5 countries is shown in Table 4.2. The table shows that non-Article 5 countries will completely phase out HCFCs by 2020, allowing 0.5% for servicing purposes during 2020–2030. Article 5 countries will complete the phase out by 2030, allowing 2.5% for servicing during 2030–2060.

TABLE 4.2
Accelerated Phase-Out Schedule of HCFCs for Article 5 and Non-Article 5 Countries

HCFC Phase-Out Dates (2013 and Beyond)	Non-Article 5 Countries (Developed Countries)	Article 5 Countries (Developing Countries)
2013	N/A	Freeze production and consumption based on the average of the 2009 and 2010 levels
2015	Reduce HCFCs by 90%	Reduce HCFCs by 10%
2020	Complete phase-out of HCFCs, allowing 0.5% for servicing purposes during the period 2020–2030	Reduce HCFCs by 35%
2025	N/A	Reduce HCFCs by 67.5%
2030	N/A	Complete phase-out of HCFCs, allowing 2.5% for servicing purposes during the period 2030–2040

TABLE 4.3
Environmental Characteristics of HCFCs, HFCs, HFC Blends, HCs, Ammonia, and CO_2

Refrigerant		Atmospheric Lifetime (Years)	Ozone Depletion Potential (ODP)	Global Warming Potential (GWP) (100 Year)
CFC (no more)	CFC-11 (Baseline ODP)	50	1	4,000
	CFC-12	102	1	10,900
HCFCs	HCFC-22	13.3	0.055	1,820
	HCFC-123	1.4	0.02	93
	HCFC-141b	9.4	0.11	630
HFCs	HFC-134a	14.6	0	1,300
	HFC-245fa	7.3	0	820
	R-32	–	0	675
HCs	HC-290 (Propane)	–	0	3
	R-1270 (Propylene)	–	0	<2
HFC Blends	R-404A	–	0	3,260
	R-407A	–	0	1,770
	R-407C	–	0	1,530
	R-410A	–	0	1,730
Ammonia	R-717	–	0	<1
CO_2	R-744	–	0	1

4.2.3 The Montréal Protocol and Hydrofluorocarbons

Hydrofluorocarbon (HFC) refrigerants have zero ODP. HFCs were used extensively as replacement of HCFCs during their phase out. However, their GWP is rather high. Table 4.3 shows the environmental characteristics of HCFCs, HFCs, HFCs blends, HCs, and two natural refrigerants: ammonia and CO_2.

Because of the high GWP of HFCs, chemical companies introduced new refrigerants with a lower GWP. Figure 4.1 shows transitional, medium-, and long-term refrigerants. There are a number of new refrigerants with a lower GWP being tested for use in HAT countries.

4.2.4 The Montréal Protocol, RTOC, and Future Refrigerants

The Ozone Secretariat for the Vienna Convention for the protection of the ozone layer and the Montréal Protocol is based at the UNEP offices in Nairobi, Kenya. Its main functions include the following:

Arranging for and serving the conferences of the parties to the Vienna Convention, meeting of the parties to the Montréal Protocol, related working groups and committees, the bureau, and the assessment panels.

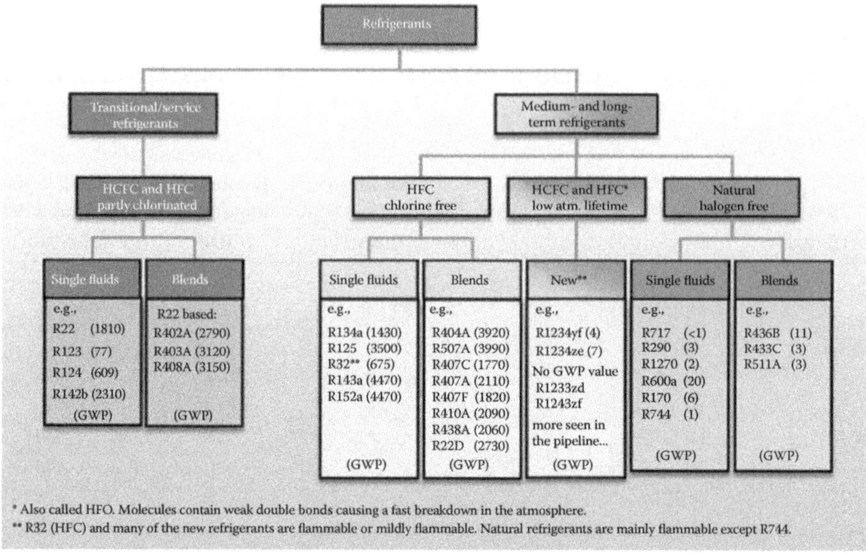

FIGURE 4.1 Transitional, medium-, and long-term refrigerants.

In 1990, the TEAP was established as the technology and economics advisory body to the MP parties. The TEAP provides, at the request of the parties, technical information related to the alternative technologies that have been investigated to help eliminate ozone depletion. The parties ask TEAP every year to evaluate technical issues related to MP, so that TEAP annual reports become a basis for decisions. TEAP manages its subsidiary bodies, Technical Options Committees (TOCs), of which the Refrigeration, Air-Conditioning and Heat Pumps Technical Options Committee (RTOC) is of special importance.

The RTOC issues a report every 4 years, written by top experts in the field, that sheds light on new state-of-the-art technologies in refrigerants, domestic appliances, commercial refrigeration, industrial systems, transport refrigeration, air-to-air air-conditioning and heat pumps, water heating heat pumps, chillers, vehicle air-conditioning, and sustainable refrigeration.

The 2014 RTOC report can be downloaded from the UNEP site:

Ozone.unep.org/sites/ozone/files/documents/RTOC-Assessment-Report-2014.pdf

The Ozone Secretariat organized the Open-Ended Working Group Meeting 36 (OEWG 36) of the parties to MP, held in Paris, France, during July 20–24, 2015.

Several proposed amendments were presented to the parties to reduce, restrict, and phase down high GWP HFC refrigerants. A decision has now been taken to phase down HFCs according to an agreement to be worked out by the stakeholders during 2016.

Table 4.4 shows the classification of refrigerants according to their 100 years GWP.[11]

TABLE 4.4
Classification of Refrigerants with 100-Year GWP Levels

100 Year GWP	Classification
<30	Ultralow or negligible
<100	Very low
<300	Low
300–1,000	Medium
>1,000	High
>3,000	Very high
>10,000	Ultrahigh

Source: RTOC Report (UNEP, 2014).

TABLE 4.5
Safety Classifications of Refrigerants

	Toxicity	
Flammability	**A**	**B**
1	A1	B1
2L	A2L	B2L
2	A2	B2
3	A3	B3

Source: RTOC Report (UNEP, 2014).

Notes: The first letter of the safety class describes the toxicity: "A" for lower level of toxicity and "B" for higher level of toxicity. The second part is the flammability of the refrigerant: 1 (no flame propagation), 2L (lower flammability), 2 (flammable), and 3 (higher flammability).

It is expected that the future refrigerants shall have a lower GWP, probably up to about 700 GWP. It is noted that lowering the GWP of refrigerants increases their flammability. This is why the majority of future refrigerants will have some flammability issues. Table 4.5 shows safety classifications of refrigerants: their flammability as well as their toxicity. A renewed interest is now being focused on natural refrigerants, such as ammonia, hydrocarbon (HC), CO_2, and water, because of their low-GWP.

4.2.5 Selecting a Suitable Refrigerant for a Chiller

In Chapter 2 of the RTOC Report (2014), a criterion is suggested for the selection of refrigerants. This is summarized as follows:

For new systems:

- Zero ODP
- Reduced climate change impact (GWP)

- Capacity and efficiency as high as possible
- Safety
- Impact on unit cost
- Availability of refrigerant commercially
- Skills and technology required for use
- Recyclability and stability

With zero ODP achieved, the remaining parameters are to be traded off against one another to achieve the optimum selection for a given application. It should be noted that almost all alternatives to HCFC-22 refrigerants have a lower coefficient of performance (COP) especially at HAT conditions. This is why there are four experimental testing programs dedicated for finding suitable lower GWP refrigerants for HAT countries.

4.2.6 Low-GWP Refrigerants for High-Ambient Temperature Countries

There are four testing programs conducted to test lower GWP refrigerants for HAT countries. These are as follows:

1. Air-Conditioning, Heating, and Refrigeration Institute (AHRI) Low-GWP Alternative Refrigerants Evaluation Program (AREP)—Low-GWP AREP for HAT countries
2. Promoting Low-GWP Alternative Refrigerants in HAT Countries for Air-Conditioning—PRAHA
3. The Egyptian Program for Promoting Low-GWP Refrigerant Alternatives—EGYPRA
4. The U.S. Department of Energy (DoE) Testing Program for Low-GWP Refrigerants for High-Ambient-Temperature Environment—U.S. DoE Oak Ridge National Laboratory (ORNL)

These evaluation programs are meant to shed light on the performance of the most commonly proposed low-GWP refrigerants and their performance degradation at HAT conditions. The objective of the programs is to assess if it is possible to achieve a similar or better energy efficiency and cooling capacity with new lower-GWP refrigerants when compared to baseline refrigerants R-22 or R-410A.

The AHRI is conducting the first program. The UNEP and the UNIDO are sponsoring programs two and three, PRAHA and EGYPRA. The U.S. DoE at ORNL is sponsoring the fourth program. The programs analyze the performance of residential split units and some central units. No program is currently looking into the performance of chillers yet; the U.S. DoE ORNL program's second phase is looking at central roof top units.

The importance of those testing programs is paramount in deciding which refrigerants are most suited to HAT conditions and will lead the way for selecting suitable future low-GWP refrigerants.

4.3 FLUOROCARBON REFRIGERANT CHILLERS

Mechanical vapor compression chillers are the most popular type of chillers for district cooling. They can be air- or water-cooled. Air-cooled chillers have capacities up to 1800 kW (500 TR). Water-cooled chillers are available at much higher capacities reaching up to 21,000 kW (6,000 TR).

Water-cooled chillers are much more efficient than air-cooled chillers. They are predominantly electric power driven. They can achieve operating efficiencies down to 0.6 kW/TR. Air-cooled chillers operate at efficiencies about 1.6 kW/TR, hence the preference of water-cooled chillers for district cooling projects.

The ideal Carnot cycle for a vapor compression chiller shows the maximum ideal COP.

Figure 4.2 shows a vapor compression cycle and a vapor absorption cycle. In this vapor compression cycle

$$COP_{ideal} = COP_{Carnot} = \frac{T_{evap.}}{T_{cond.} - T_{evap.}}$$

where:

COP_{ideal} is the Carnot COP
$T_{evap.}$ is the evaporation temperature, °R (K)
$T_{cond.}$ is the condensing temperature, °R (K)
COP_{Carnot} is the limit of COP in an ideal cycle

In an actual cycle, the built-in inefficiencies reduce COP, and they vary from 4 to 6.5 for an electric chiller and correspond to 0.88–0.54 kW/TR.

There are three predominant types of electric chiller in district cooling systems: reciprocating, screw, and centrifugal. Mechanical vapor compression chillers can also be engine or turbine driven.

Many electric chillers can be obtained that will operate on low, medium, or high voltages.

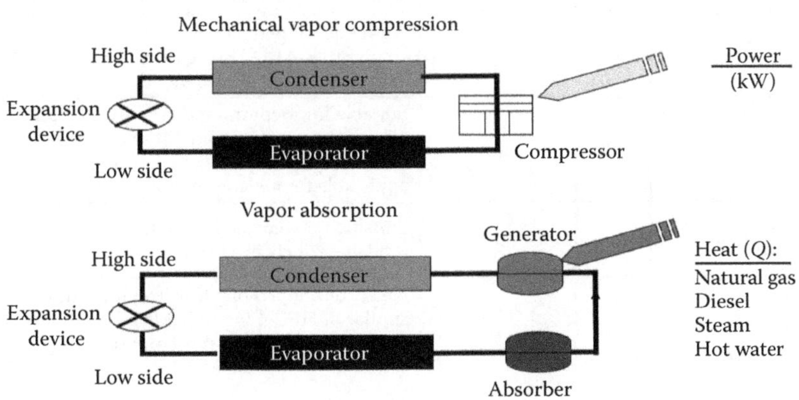

FIGURE 4.2 Schematic diagrams of a vapor compression cycle and a vapor absorption cycle.

4.3.1 Centrifugal Chillers

Electric power-driven centrifugal chillers are the dominant type of mechanical vapor compression chillers in district cooling systems. One of the most important characteristics of this type of chiller is its ability to operate down to 3°C. Centrifugal chillers have a capacity range of 1,000–21,000 kW (300–6,000 TR), which makes them suitable for various district cooling projects. They are usually cooled by condenser water from cooling towers and can operate at a condenser-entering water temperature up to 37°C.

MP and its amendments have changed the perception of refrigerants that are suitable for chillers, especially since centrifugal chillers can operate usefully for more than 20 years.

This is why Section 4.2 dealt with the impact of MP on future refrigerants. It is recommended that before selecting refrigerants for chillers, one should look into that section to find which refrigerant is most suited. Figure 4.3 shows the new alternative low-GWP refrigerant replacing HFC refrigerants for chillers.

Figure 4.4 shows a centrifugal water-cooled chiller that has a nominal capacity of 10,560 kW (3000 TR) with two compressors.

Refrigerant	GWP	Flammability[2]	Comments
Chillers with Centrifugal Compressors			
HFO-1234ze	7	2L	Being developed for large centrifugal chillers as an alternative to HFC-134a.
HFO-1233zd HFO-1336mzz	5 9	1 1	New fluids suitable for low pressure centrifugal chillers as and alternative to HCFC-123.
R-718 (water)	0	1	Water can be used as a refrigerant in chiller systems, but requires very large compressor swept volume.
Chillers with Positive Displacement Compressors			
HFO-1234ze	7	2L	Already available in a range of small and medium-sized chillers.
HFO-32	675	2L	Has performance similar to R-410A and is suitable for small and medium-sized chillers.
R-446A R-447A	460 582	2L 2L	Newly developed blends with properties similar to R-410A. Being considered for small and medium-sized chillers.
R-717 (ammonia)	0	2L	Suitable for medium-and large-sized chillers with screw compressors. More commonly used for industrial chillers but can also be applicable for air-conditioning.
HC-290 HC-1270	3 2	3 3	Suitable for small and medium-sized chillers, available widely in Europe.
R-450A R-513A	601 631	1 1	Newly developed blends with properties similar to HFC-134a, suitable for medium-sized chillers using screw compressors.

FIGURE 4.3 Alternative refrigerants to HFCs for chillers with low-GWP. (By kind permission of Daikin Middle East & Africa.)

FIGURE 4.4 Centrifugal water-cooled chiller (3000 TR). (Model WCT, All rights reserved. Copyright 2016 Daikin Middle East & Africa.)

4.4 NONFLUOROCARBON REFRIGERANT CHILLERS

4.4.1 Absorption Chillers

4.4.1.1 Absorption Theory

One of the oldest methods of creating cold is absorption refrigeration; Ferdinand Carré in 1858 is credited with inventing a continuous absorption refrigerator. Absorption refrigeration chillers today represent an important percentage of the total number of chillers produced worldwide, especially in China, Japan, South Korea, India, Pakistan, and other Asian countries.

Unlike vapor compression, an absorption machine does not utilize a compressor and motor. Instead, two heat exchangers perform this task. Those are the absorber and the generator.

Figure 4.5 shows a single-effect (one generator) absorption cycle. Thus, the main advantage of absorption refrigeration technology is that it is heat driven, not work driven, as is the case for vapor compression.

Absorption machines are operated by applying heat to the generator in the form of direct or indirect-fired heat; those are the two known forms of absorption technology. In direct-fired technology, generators are equipped with burners, which are fired by natural gas, liquefied petroleum gas (LPG), light oil, heavy oil, and other liquid, or gaseous fuels. Indirect-fired chillers are driven by steam, hot water, or exhaust heat exchangers.

Figure 4.6 shows a single-effect absorption cycle.[12] The cycle tilts to the right to simulate the log p versus $1/T$ vapor pressure diagram of a lithium bromide-water absorption solution, the Dühring diagram of the mixture. The Dühring diagram of the solution is shown in Figure 4.7 for various concentrations of $LiBr\text{-}H_2O$ solutions.

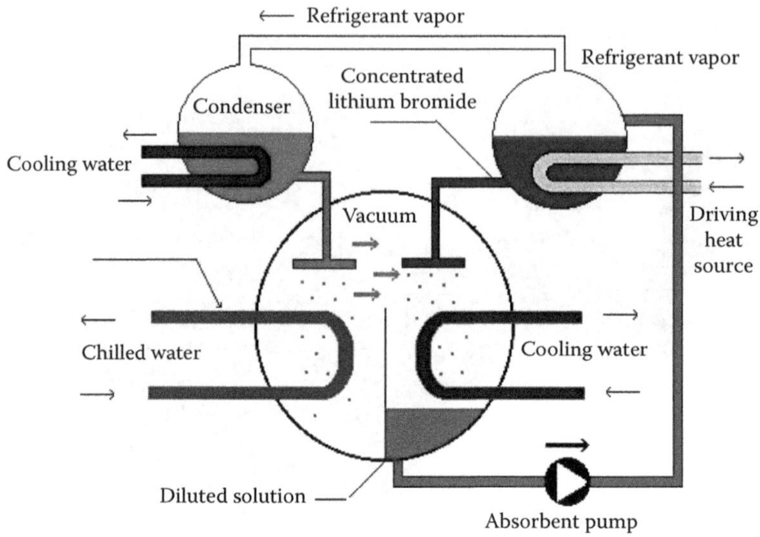

FIGURE 4.5 Schematic diagram of a single-effect (one generator) absorption chiller.

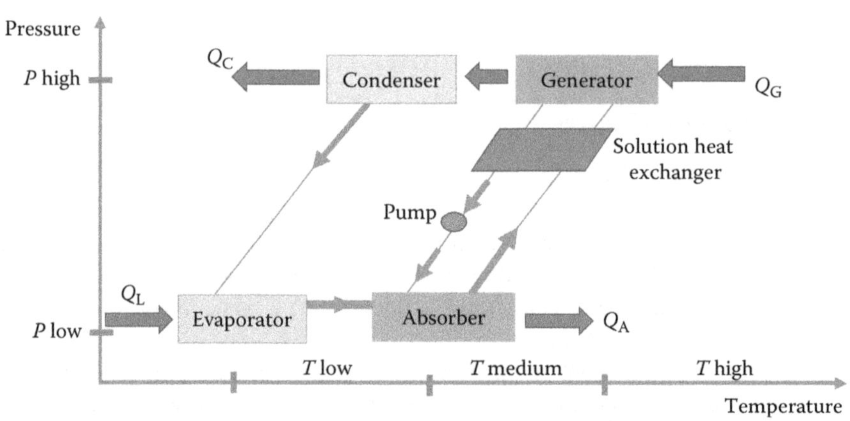

FIGURE 4.6 Schematic diagram for a single-effect absorption cycle.

Figure 4.8 shows a single-effect cycle superimposed on a Dühring diagram illustrating the various stages of the cycle.

The absorption cycle utilizes an absorbent in addition to a refrigerant. This pair works together as a solution in the right part of the cycle, whereas in the left part of the cycle only the refrigerant exists. The function of the absorbent is to absorb the refrigerant vapor exiting the evaporator and depress its total pressure. Together it mixes with the absorbent and forms a dilute solution (dilute in the absorbent), which is pumped through the solution heat exchanger where it absorbs heat from the concentrated solution exiting the generator and help reduce the concentrated solution temperature, before reaching the generator.

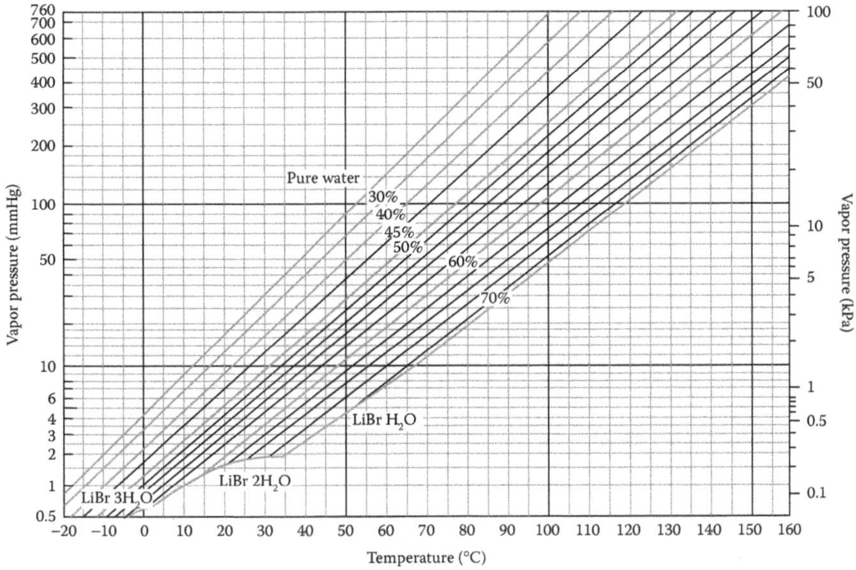

FIGURE 4.7 Dühring chart for LiBr solutions.

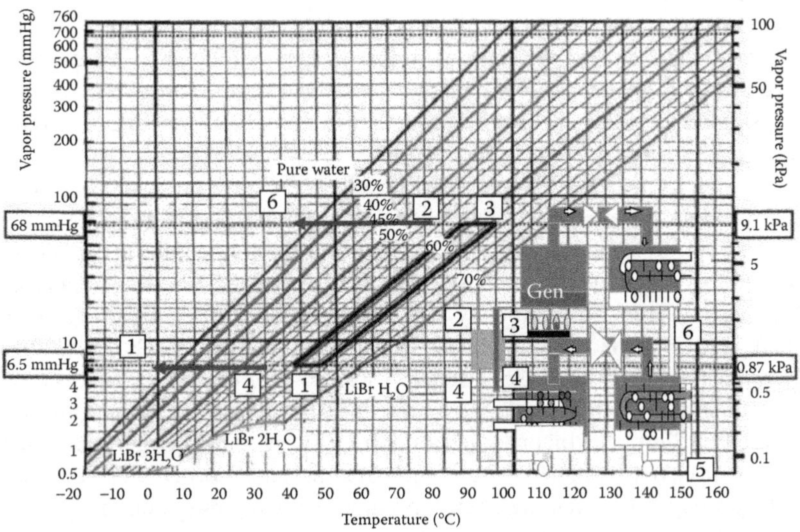

FIGURE 4.8 Dühring chart for an LiBr solution with a typical single-effect cycle flow.

In the generator, heat is applied and the refrigerant is released and passed to the condenser. There are many absorption-working solutions; $LiBr/ZnBr_2$ (2/1 mol) and methanol have been researched as possible solutions[13] among other mixtures. The most famous solution pairs are as follows:

The absorbent in both solutions lithium bromide-water and water-ammonia is the first term in the binary mixture, whereas the second is the refrigerant.

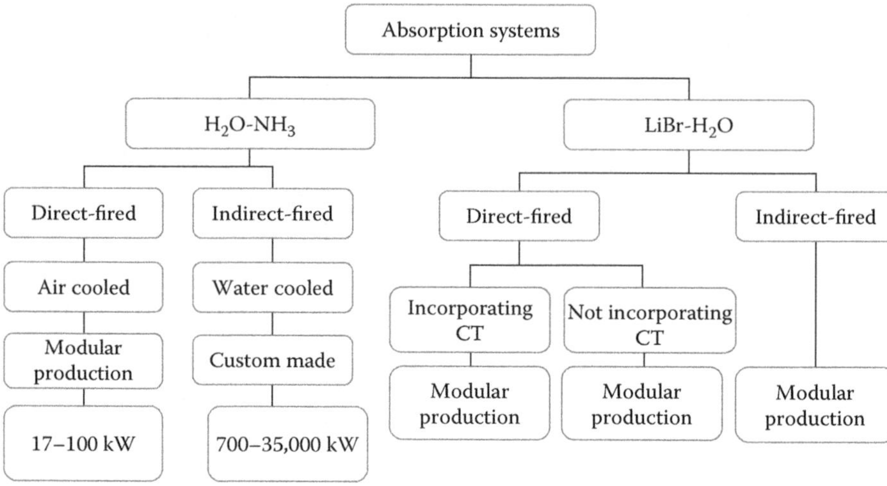

FIGURE 4.9 The current refrigeration capacities available for LiBr-H_2O and H_2O-NH_3 absorption systems.

Figure 4.9 shows the current refrigeration capacities available for both established absorption systems.

Absorption systems using water-ammonia working solutions are used for low-temperature refrigeration applications as well as for air-conditioning, because ammonia, a refrigerant, can operate at low evaporation temperatures. Lithium bromide-water working solutions operate at above freezing temperatures; water is the refrigerant. They are used only for air-conditioning applications.

Small capacity water-ammonia absorption systems are air-cooled, produced in modular production, in capacities and multiples of 10–17 kW (3–5 TR); they are used in air-conditioning. Larger capacity water-ammonia systems are made to order with capacities from 700 to 35,000 kW (200–10,000 TR) and are water-cooled. Those larger systems are primarily used for process cooling in food, beverages, chemical, and pharmaceutical plants. Waste heat, which is readily available in those applications, drives those systems. Lithium bromide-water absorption units are more widely used than water-ammonia units. They are direct- or indirect-fired. Direct-fired chillers can incorporate a cooling tower with refrigeration capacities from 17 to 100 kW (5–30 TR). These units are shell and coil types of construction, unlike larger capacities that are shell- and tube types of construction.

Lithium bromide chillers are all water-cooled. The larger capacity direct-fired chillers are common in Asia and parts of the Middle East. They range in refrigeration capacities from 100 to 4200 kW (30–120 TR). Those chillers are used in district cooling applications. Indirect-fired absorption chillers operating with lithium bromide-water working solutions are used where waste heat is available. Those chillers utilize waste heat in the form of recovered hot water or steam. They are also used where boilers must run round the year such as in hospitals. Other applications include cogeneration systems where the waste engine's heat or steam, or turbine exhaust is available. Some lithium bromide-water absorption

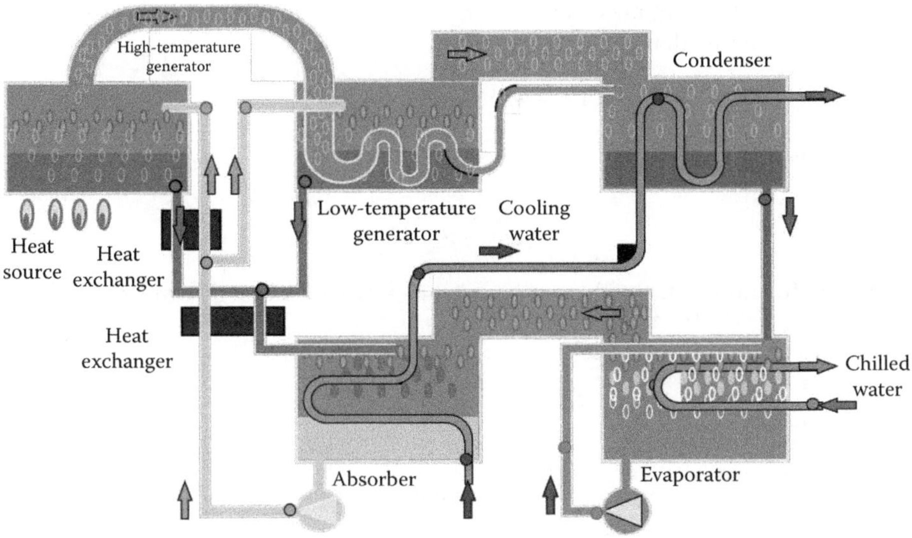

FIGURE 4.10 Schematic diagram of a double-effect direct-fired absorption chiller cycle.

chillers are constructed so that they operate directly by waste flue gases, although the waste flue gas stream may be corrosive and quickly damage the generator part of the chiller.

Currently, lithium bromide-water absorption chillers of the double-effect type (two generators) are most common. Figure 4.10 shows a double-effect direct-fired absorption cycle. In this cycle, the dilute solution exiting the absorber is pumped in parallel to both high- and low-temperature generators.

The heat source is applied to the high-temperature generator, and the ensuing vapor heats the low-temperature generation. Other pumping schemes for the absorber to generator solution exist, such as in-series, reverse parallel, or other arrangements. Double-effect absorption units have higher COPs than single-effect absorption units, 0.7–0.8 for single effect and 1.2–1.35 for double effect.

LiBr-H_2O mixtures are not totally miscible across the whole concentration spectrum. LiBr crystallizes at certain concentrations forming crystals of LiBr.5H_2O, LiBr.3H_2O, LiBr.2H_2O, and LiBr.H_2O. Calculations of thermodynamic state point of solutions require knowledge of concentration calculations.[14]

When two mixtures are mixed, a third mixture results with a different concentration. The composition is expressed as the mass fraction of one of the components. The mass fraction of lithium bromide is expressed here as ξ. In a solution, *mL*, is the mass of LiBr and *mw* is the mass of water. The mass fraction of LiBr is

$$\xi = \frac{mL}{mL + mw}$$

The mass fraction of H_2O is

$$1 - \xi = \frac{mw}{mL + mw}$$

Mixing two solutions of masses m_1 and m_2 having mass fractions ξ_1 and ξ_2 produces a solution of intermediate mass fraction ξ_3.

$$(m_1 + m_2)\xi = m_1\xi_1 + m_2\xi_2$$

In addition

$$\frac{m_2}{m_1} = \frac{\xi_3 - \xi_1}{\xi_2 - \xi_3}$$

The mole fraction is denoted by ψ. The number of moles in a mixture of LiBr and H_2O is

$$nL = \frac{mL}{ML}$$

In addition

$$nw = \frac{mw}{Mw}$$

where ML and Mw are the molar masses (molecular weight) of LiBr and H_2O, respectively. The mole fractions of LiBr is ψ:

$$\psi = \frac{nL}{nL + nw}$$

In addition

$$1 - \psi = \frac{nw}{nL + nw}$$

Figure 4.11 shows both the Dühring diagram of mixtures of LiBr-H_2O and the enthalpy versus concentration diagram for mixtures.

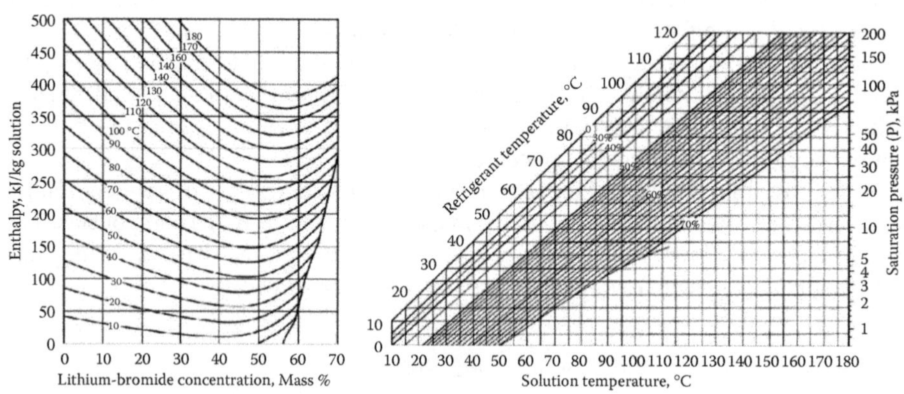

FIGURE 4.11 Dühring diagram and enthalpy versus a concentration diagram of LiBr-H_2O mixtures.

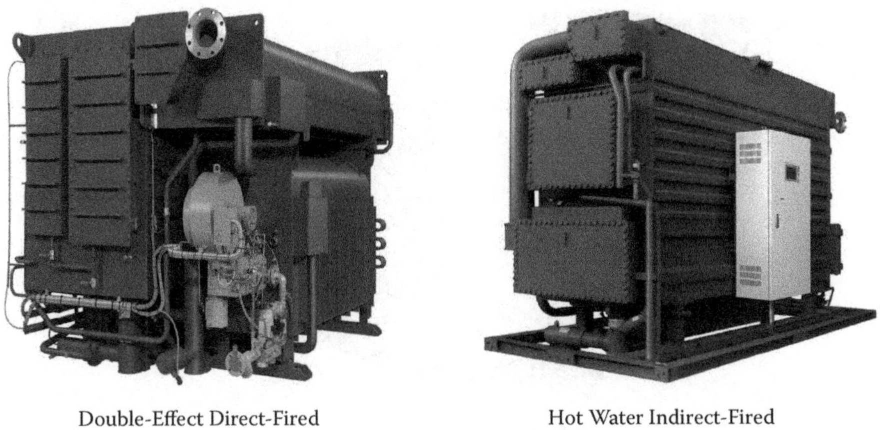

FIGURE 4.12 Absorption chillers, double-effect direct-fired and hot water indirect-fired. (By kind permission of Kawasaki Thermal Engineering, Japan.)

Figure 4.12 shows absorption chillers: a double-effect direct-fired and hot water indirect-fired chillers.

Figure 4.13 shows absorption chillers: steam-fired, hybrid (gas and hot water-fired), and a triple-effect unit. Triple-effect chillers are available commercially. These chillers have three generators, and their COPs are approximately 1.85 higher than those of double-effect chillers.

4.4.1.2 Absorption: Historical Perspective

In the early 1960s, absorption chillers were first manufactured in the United States. The four major U.S. air-conditioning manufacturers produced absorption chillers in quantities comparable to vapor compression chillers. The gas regulation laws changed the picture dramatically. This was the result of the "cold war" in anticipation of a conflict between the super powers and saving oil and gas resources in the United States. Figure 4.14 shows absorption chiller shipments in the United States from 1965 to 2000.[15]

Absorption technology depended at the time on the continuous exploration of oil and gas resources. The decline persisted until it reached a low in the 1980s of 20,000–30,000 TR/year from a high of more than 500,000 TR/year. In the 1990s with deregulation, production increased to about 200,000 TR/year.

Japan started manufacturing absorption chillers, and direct-fired chillers were produced in the late 1960s. Figure 4.15 shows worldwide absorption unit production reached 6550 units in 2005. Today, China is the largest producer of absorption chillers.

Figure 4.16 shows the rest of the world market, excluding Japan, China, South Korea, and the United States, produced 860 chillers in 2005.

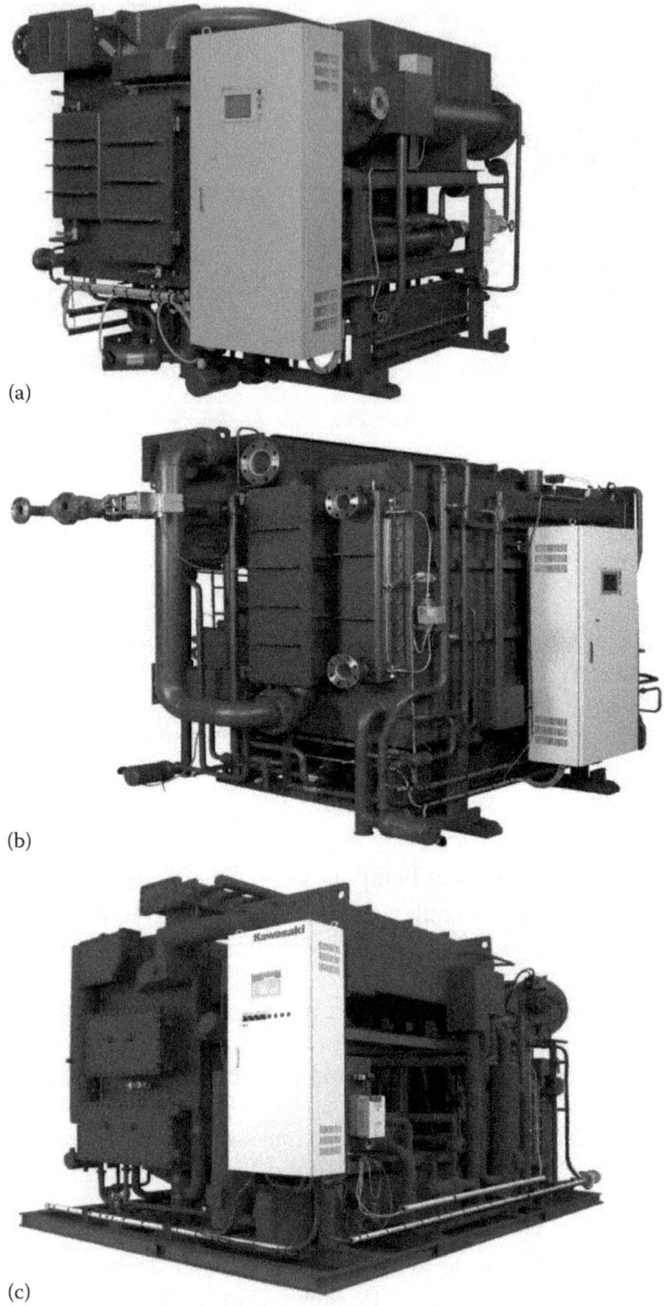

FIGURE 4.13 Absorption chillers: (a) steam-fired, (b) hybrid (gas and hot water), and (c) triple effect. (By kind permission from Kawasaki Thermal Engineering, Japan.)

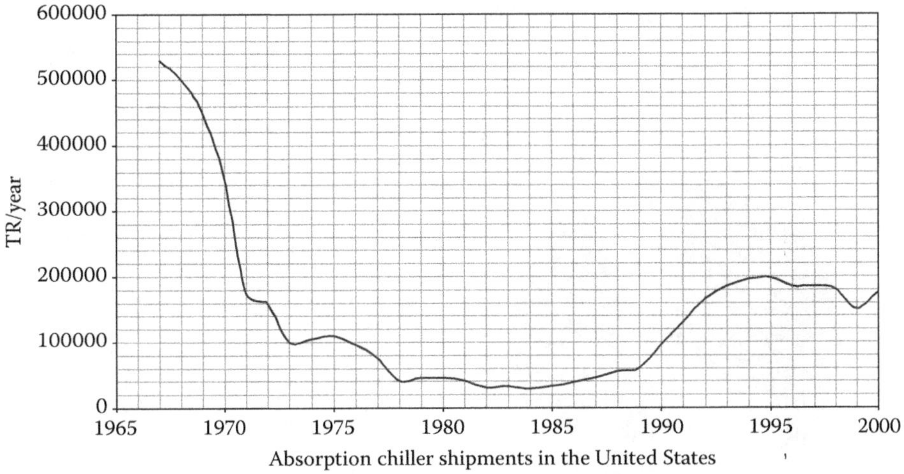

FIGURE 4.14 Domestic shipments of large tonnage absorption chillers in the United States, 1965–2000.

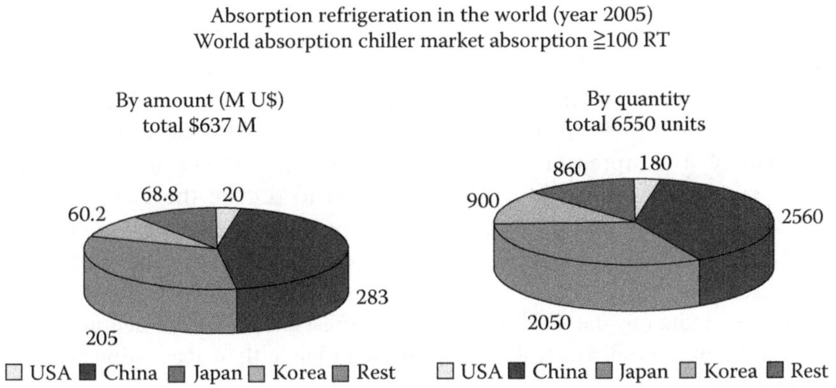

FIGURE 4.15 Absorption production worldwide in 2005.

4.4.1.3 COP (Heat Ratio) Absorption versus COP Vapor Compression

Figure 4.17 shows the development of the COP (heat ratio) of absorption chillers from 1968 until 2005. In this graph, the COP development timeline is shown together with the various political and economic milestones that precipitated change in Asia.

The graph shows the development of both absorption chiller categories; less than and over 350 kW (100 TR). The COP of more than 350 kW (100 TR) absorption chillers improved from 0.75 in 1968 to 1.35 in 2000.

For less than 350 kW absorption chillers, the COP improved from 0.47 in 1968 to 1.2 in 1990. Triple-effect chillers have a COP of 1.68–1.85.

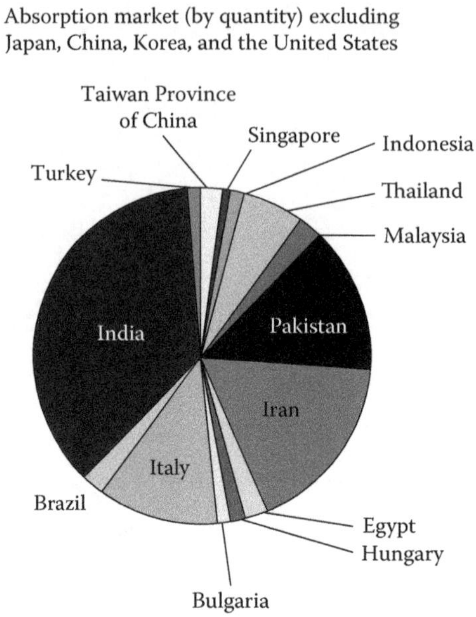

Country	Quantity
Taiwan Province of China	20
Singapore	5
Indonesia	10
Thailand	50
Malaysia	20
Pakistan	120
Iran	150
Egypt	20
Hungary	10
Bulgaria	10
Italy	100
Brazil	20
India	315
Turkey	10

FIGURE 4.16 The rest of the world market in 2005.

It is important to note that the COP of an absorption chiller is a heat ratio and cannot be compared directly with that of a vapor compression chiller, because the latter is a ratio of useful refrigeration effect divided by the work of an electric motor-driven compressor, a different, more refined form of energy exergy. In order to compare both COPs, it is important to reduce the COP of a vapor compression chiller using the inefficiencies of producing electric power, such as the electric power plant efficiency, transformers efficiencies, and transmission efficiencies.

At the end, reducing the COP of vapor compression to 35% is nearer its actual value. Vapor compression systems are thus more efficient than absorption systems by a small margin.

However, the uniqueness of absorption systems in utilizing a crude form of energy and their low electric power requirement make them an important element when choosing a chiller for a district cooling plant.[16]

4.5 DC CHILLER PLANT ARRANGEMENTS

There are two chiller arrangements: parallel and in-series. Both arrangements are used with centrifugal, reciprocating, screw, and absorption chillers, or a combination of them.

Figure 4.18 shows a schematic diagram of parallel and in-series flow chiller arrangements.

Series flow chiller arrangements are used when a large ΔT is required for the chilled water, with an increased chiller efficiency. The pressure differential is

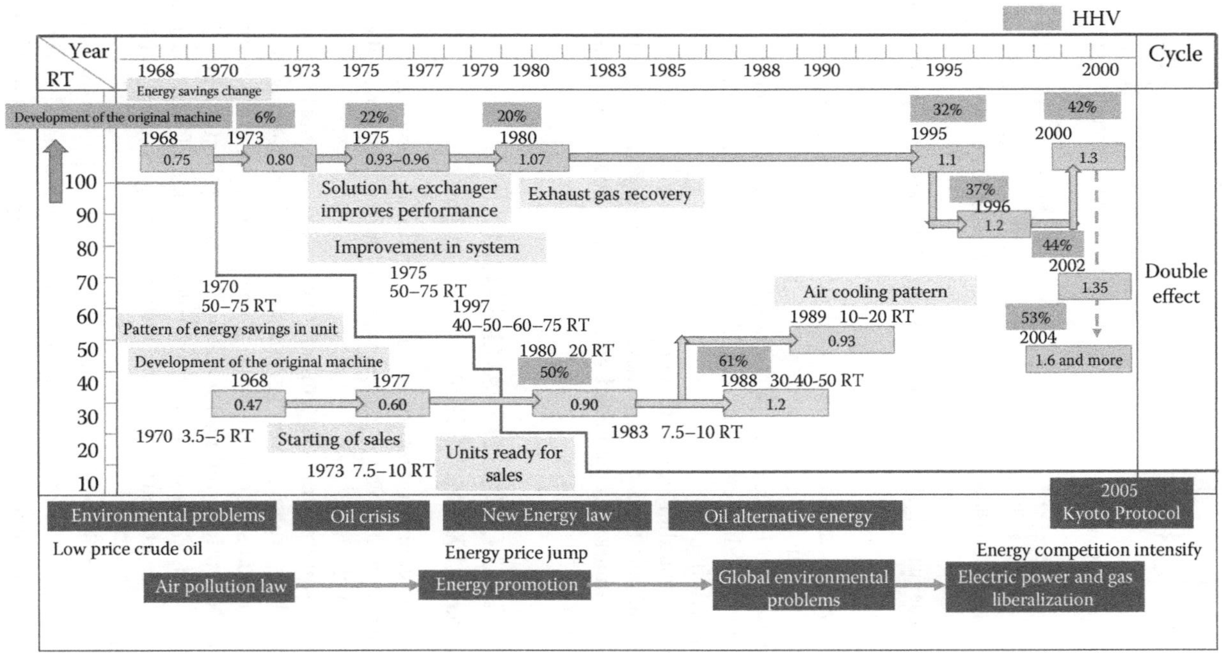

FIGURE 4.17 The development of COP of absorption technology in Japan—from 1968 to 2000.

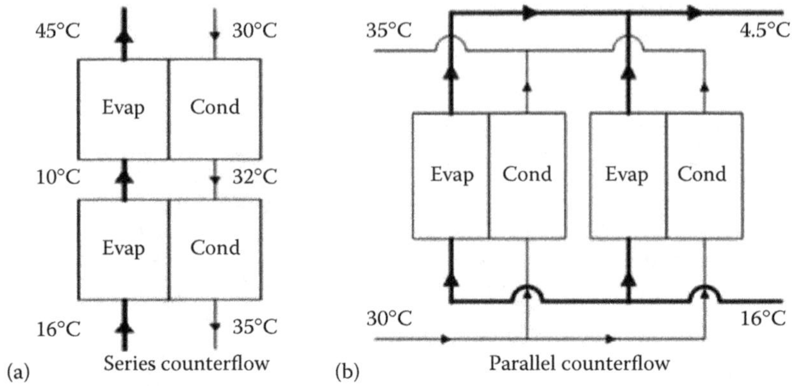

FIGURE 4.18 Schematic diagram of (a) in-series flow and (b) parallel flow chiller arrangement.

however higher when compared to parallel flow chillers. This may lead to a larger pumping power consumption.

With parallel flow chiller arrangements, smaller pressure differentials occur. Also with a primary pump dedicated to each chiller, failure of pump will not cause the two chillers to stop operating.

However, with a parallel flow arrangement, it is easier to operate the plant at a higher efficiency at part load because there is less refrigeration capacity to control. This is despite the smaller ΔT achieved with a parallel flow arrangement compared to an in-series flow.

4.6 GENERATING CHILLED WATER FROM RECOVERED EXHAUST HEAT

The highest possible economic return is achieved by heat recovery, especially when comparing electric operated vapor compression chillers with recovered exhaust heat operated absorption chillers.[17]

This system is suitable for use with recovered exhaust heat from ovens, furnaces, and reject hot streams from various industrial operations. The recovered heat is used to drive absorption chillers of the lithium bromide-water type, and sometimes in the chemical and petrochemical industry, water-ammonia absorption chillers are used. Both direct and indirect methods of firing are used with LiBr-H_2O absorption chillers, whereas indirect firing is used with H_2O-NH_3 absorption chillers.

The chemical composition of the source heat stream is an important factor in deciding which method to use to recover exhaust heat. A clean exhaust makes it possible to utilize direct exhaust-fired absorption chillers. A dirty exhaust containing harmful material is indirectly utilized to produce steam or hot water, which are then used to fire indirect-fired absorption chillers. A gaseous heat stream from an oven or furnace would typically require a gas-to-steam or gas-to-water heat recovery unit.

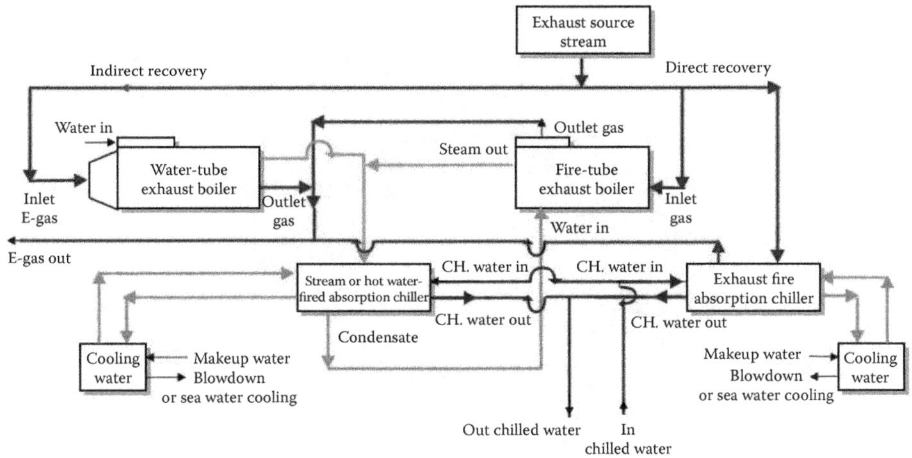

FIGURE 4.19 Schematic diagram of chilled water production by recovered exhaust heat.

Figure 4.19 shows a schematic diagram of a recovered exhaust heat to produce chilled water. In the diagram, both water tube and fire tube exhausts are used, as well as direct-fired exhaust absortion chillers.

Exhaust-fired systems are less costly than both water tube and fire tube systems, although their useful life expectancy is shorter. Both water tube and fire tube exhaust are used with the ease of cleaning the tube; the major factor is selecting which system.

4.7 DISTRIBUTED DISTRICT COOLING STATIONS

In district cooling applications where it is not economically advantageous to build the chilled water distribution network at an early stage, it may be beneficial to locate generic DC plants near the building or buildings that need the service. This is also true of remote locations for a chilled water distribution network when the envisaged refrigeration capacity has not been provided because of a late load addition to the master plan. In this case and similar cases, when new DC plant capacity needs to be added to a DC system, the concept of a modular central plant (MCP) becomes a useful solution.

A MCP combines chillers, cooling towers, pumps, and piping in a factory engineered and assembled package. This concept also reduces site activities and thus reduces cost and assembly time, which leads to faster delivery of chilled water.

Figures 4.20 and 4.21 both show a view inside an MCP and a MCP with cooling towers installed on top of it. The MCP can become a permanent part of the distributed district cooling system or may be dismounted in a traditional centralized district cooling system when the chilled water network is extended to the buildings it serves and is then used in other locations where needed.

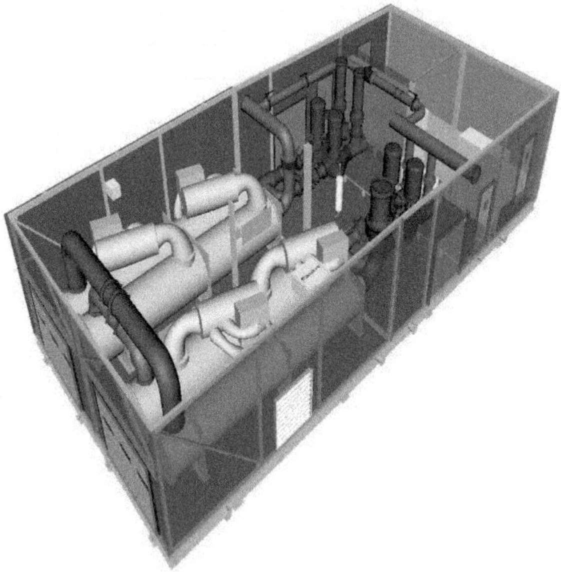

FIGURE 4.20 View of the inside of a modular central plant for distributed district cooling. (Modular chiller plant concept design. All rights reserved. Copyright 2016 Daikin Middle East & Africa.)

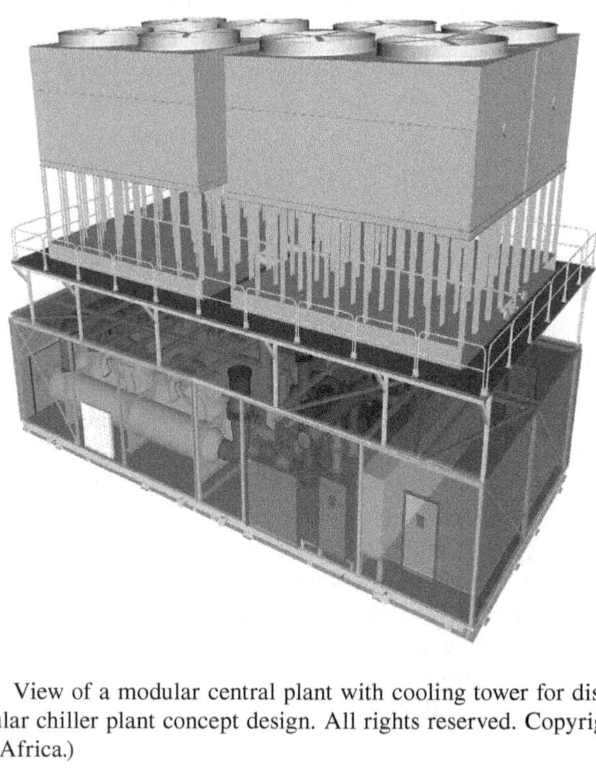

FIGURE 4.21 View of a modular central plant with cooling tower for distributed district cooling. (Modular chiller plant concept design. All rights reserved. Copyright 2016 Daikin Middle East & Africa.)

5 Designing Chilled Water Distribution Systems

5.1 CHILLED WATER PUMPING ARRANGEMENTS

5.1.1 Constant Flow Arrangements

The design of chilled water distribution systems is an important aspect of district cooling (DC) design, because it is one of the three vital parts of the system together with chilled water production and chilled water utilization by end users. There are two types of chilled water distribution: constant flow and variable flow.

Figure 5.1 shows a constant flow primary–secondary pumping scheme. Constant flow arrangement systems are applied to small capacity district cooling systems where the advantages of variable flow systems are not appreciable. In this pumping scheme, both in-series and parallel chiller arrangements can be used, although chillers arranged in series will provide larger ΔTs across supply and return chilled water lines, thus reducing the size and cost of the distribution system.

Parallel chiller arrangement will produce reduced chilled water supply temperatures, especially at the part load. It may be necessary then to operate more chillers to maintain flow rates, although chillers will run at part load conditions with reduced efficiencies.

5.1.2 Variable Flow Arrangements

The primary advantages of these arrangements are their reduced consumption of pumping energy and use of distribution system diversity, saving pumping energy. These systems are used in relatively larger DC systems. There are four main variable flow-pumping arrangements:

1. Variable speed primary pumping
2. Primary–secondary pumping
3. Primary–secondary–tertiary pumping
4. Primary-distributed secondary pumping

5.1.2.1 Variable Speed Primary Pumping

Figure 5.2 shows a variable flow primary pumping arrangement. In this system, the primary pumping regulates the chilled water flow according to the load demand. Pumping energy consumption is reduced compared to constant speed. This system is suitable when the plant pumps can satisfy buildings' pressure drops, otherwise buildings with larger pressure drops may not be served adequately. Monitoring building

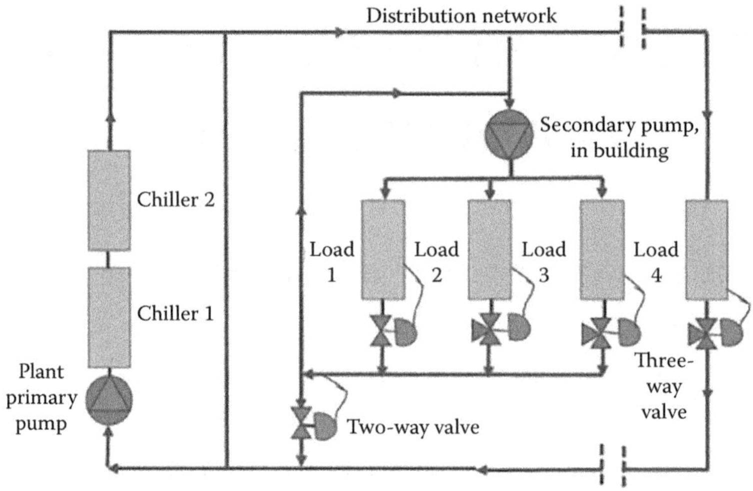

FIGURE 5.1 Constant flow primary–secondary pumping arrangement.

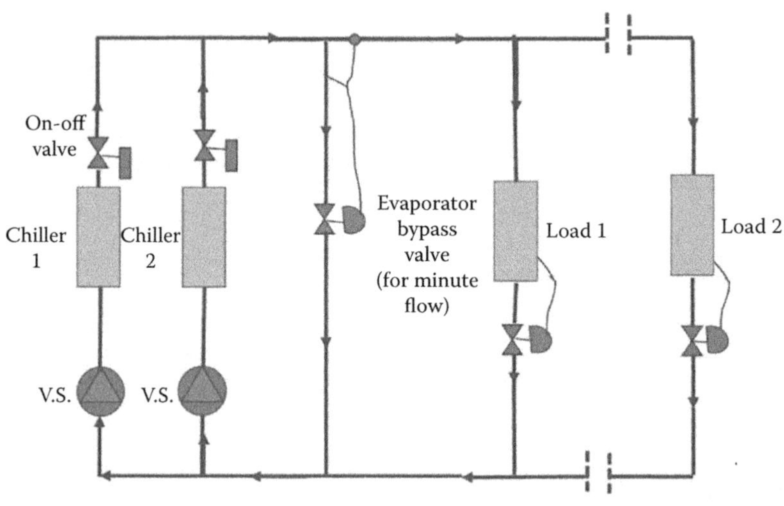

FIGURE 5.2 Variable flow primary pumping arrangement.

pressure drops at the plant is essential to operate primary pumps at speeds providing necessary flow rates and pressure drops.

Figure 5.3 shows a pressure gradient diagram at full and partial loads along the various distances of buildings served by the distribution system. These diagrams are useful to find out the pressure and flow of the pumping arrangement, thus checking pump energy consumption. Note at the part load condition how the supply and return pressure gradients are moving upward and downward, respectively, thus creating a larger Δp at the end of these lines.

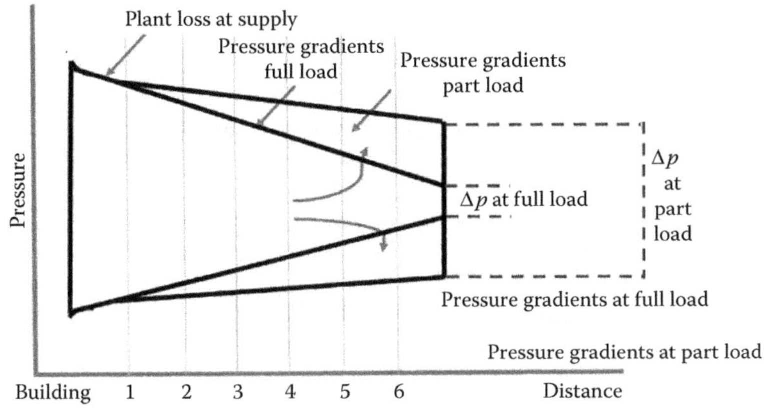

FIGURE 5.3 Pressure gradient diagram at full and partial loads.

5.1.2.2 Primary–Secondary Pumping Arrangements

Figure 5.4 shows a primary–secondary pumping arrangement, and Figure 5.5 shows the primary–secondary pumping pressure gradients. This system is used when the district chilled water distribution system is long, and the variable primary system cannot cope with flows and pressure drops. This arrangement is flexible when the district expansion scheme is not clear at inception, and additional buildings may be added at a later stage.

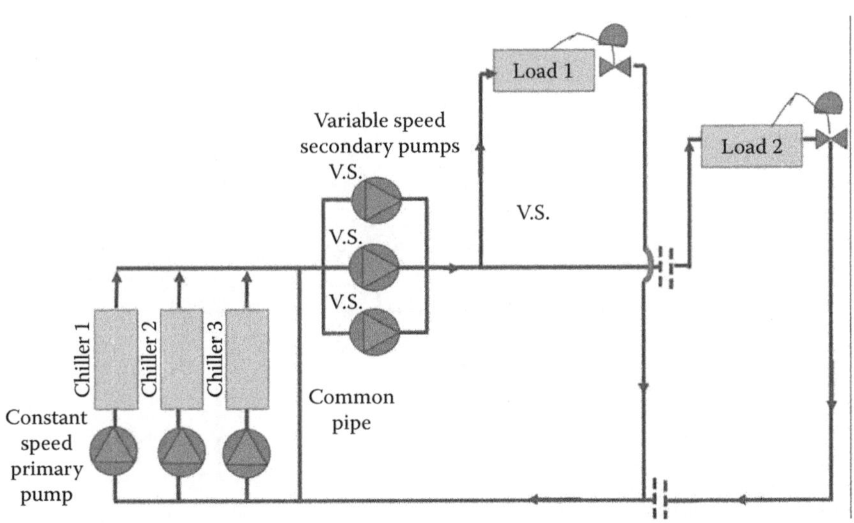

FIGURE 5.4 Primary–secondary pumping arrangement.

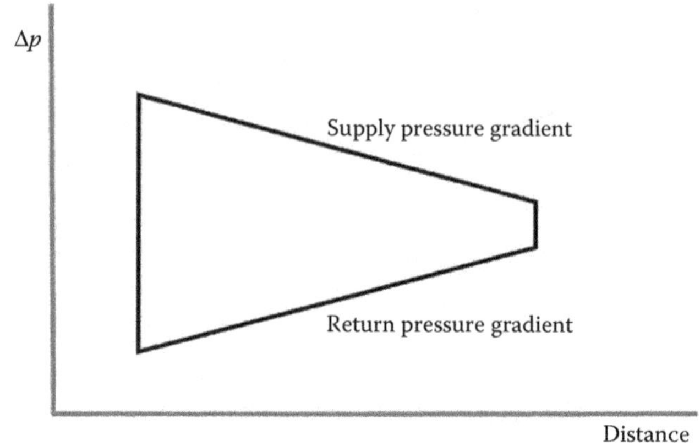

FIGURE 5.5 Primary–secondary pumping pressure gradients.

5.1.2.3 Primary–Secondary–Tertiary Pumping Arrangements

It may be necessary when supply and return chilled water distribution lines become too long to add in-building pumps to provide the necessary flow and pressure for each building.

Figure 5.6 shows a primary–secondary–tertiary pumping arrangement, and Figure 5.7 shows primary–secondary–tertiary pressure gradients. These systems are also used in districts with heavy loads in buildings.

5.1.2.4 Primary–Secondary Distributed Pumping Arrangements

Some systems may have a very large cooling load. It is possible for this system to use a primary–secondary distributed pumping arrangement. Figure 5.8 shows a

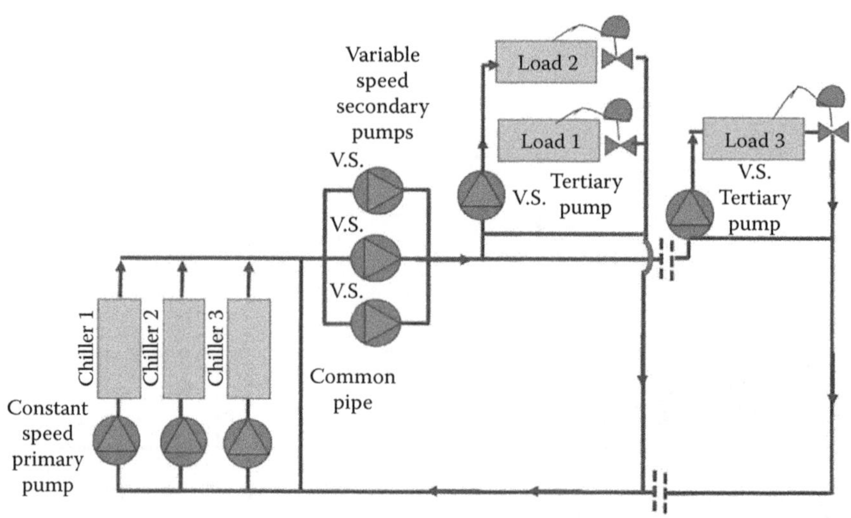

FIGURE 5.6 Primary–secondary–tertiary pumping arrangement.

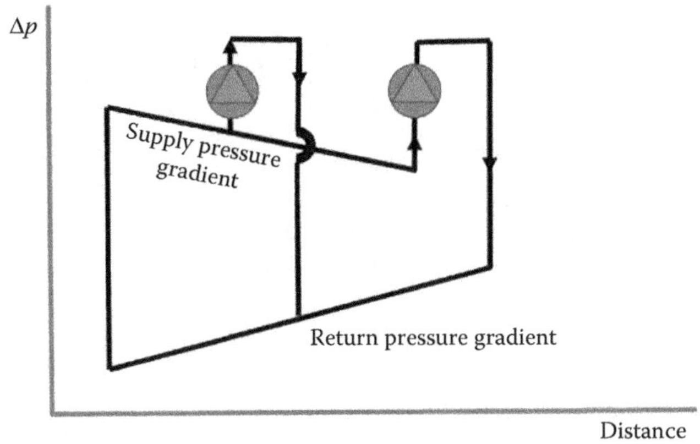

FIGURE 5.7 Primary–secondary–tertiary pumping pressure gradients.

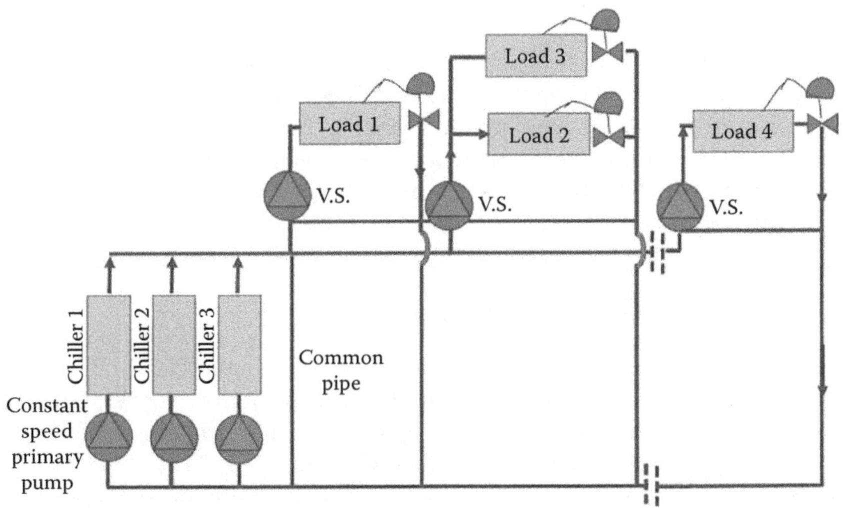

FIGURE 5.8 Primary–secondary distributed pumping arrangement.

primary–secondary distributed pumping arrangement. Figure 5.9 shows a primary–secondary distributed pressure gradients.

This system is probably the most suited system for large districts, because it eliminates secondary pumps in central plants. A reduction in total chilled water pump power of 20%–25% is possible. Although this system is highly attractive, it is not suitable for districts where additional buildings may be added at a later stage. The chilled water supply gradient pressure is lower than the return gradient in these systems. Pipes are oversized compared to other systems, which increases the initial capital cost. The operational savings mitigate all those factors in large systems.

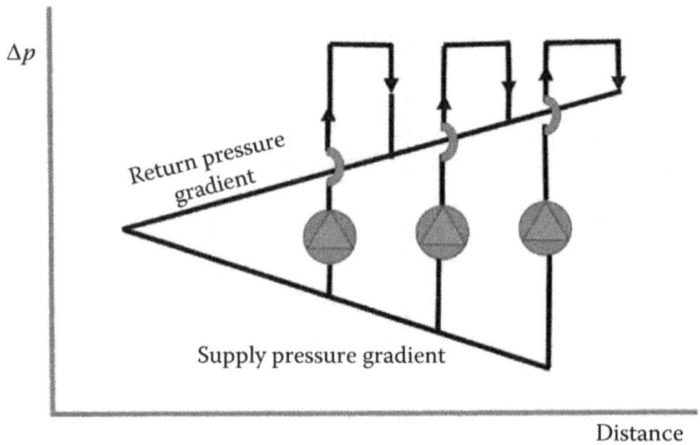

FIGURE 5.9 Primary–secondary distributed pressure gradients.

Because the supply gradient is low in those systems, maintaining a pressure at the plant higher than the net positive suction head (NPSH) becomes an important issue. This is done by reducing the speed of chilled water in the piping, thus reducing the pressure drop and sometimes pressurizing the expansion tank/s.

It is important in the arrangement that the plant operator can monitor and control pressures at the user buildings. This will allow the operator to provide an effective, efficient, and economical response to changing loads.

5.2 PIPING NETWORK MATERIAL

Chilled water pipes may be made from a variety of piping materials. These are as follows: carbon steel, HDPE (high-density polyethylene), iron, copper, GRP (glass reinforced plastic)/GRE (glass reinforced epoxy), cement (including reinforced concrete and polymer mortar), and PVC (polyvinylchloride).

These piping materials vary in their properties and installation procedures.

ASHRAE *District Cooling Guide*[18] (©ASHRAE www.ashrae.org, 2013, chapter 4) summarizes the relative merits of piping systems.

Table 5.1 has been prepared using materials published in the guide. It shows the advantages and disadvantages of various piping materials and the standards used.

Figure 5.10 shows the relative cost of piping alone (uninsulated). This figure shows that steel pipes are competitive up to a diameter of 350 mm (14 in.). More than this size, prices of steel pipes tend to increase dramatically. Piping size diameters between 460 (18 in.) and 1140 mm (45 in.) of steel pipes are several times more expensive when compared to concrete or PVC pipes. Ductile iron pipes (DIP) and HDPE pipes are also cost competitive.

Many types of piping materials have been used in district cooling applications, as shown in Table 5.1.

TABLE 5.1
Merits of Piping Materials of DC Piping Networks

No.	Type	Advantages	Disadvantages	Standards
1	Seamless/ welded carbon steel	• High strength • Flexible • Can achieve good joint integrity by welding • Familiar material to workforces • Available in sizes up to 1500 mm (60 in.)	• Relatively expensive compared to other material • Requires corrosion protection • Requires cathodic protection when directly buried • Requires skilled labor force • Slower installation in large diameters	• ASTM A53/ A53M (2012a) • ASTM • A106A/106M (2011), C200 (2012)
2	Ductile iron	• Moderate strength and flexibility • Familiar material to workforces • Faster installation • Available from 100 to 1600 mm (4–64 in.)	• Heavy • Corrodes, needs corrosion protection • Joining by mechanical systems only • Expensive fittings • Allowable leakage per standards	• AWWA C 151 (2009)
3	GRP/GRE	• Lightweight • High strength in all sizes	• Poor flexibility • Joining by cement/ field layup or mechanical joint • Difficult to add branch • Unfamiliar material with many workforce • Slow to cure in low ambient temperatures • Finding leakage point may be difficult	• AWWA C 905 (2013) • ASTM D2996 (2007) e 1
4	Cement pipes (include reinforced concrete and polymer mortar)	• Reasonable strength • Available in all sizes • Familiar material with many workforces	• Heavy, poor flexibility • Mechanical joints only • Thrust blocks required • Adding branch difficult • Limits on pressure and velocity • Allowable leakage per standards	• AWWA C 300 (2011a) • AWWA C 301 (2007a) • AWWA C 302 (2011b) • AWWA C 303 (2008a)

(*Continued*)

TABLE 5.1 (*Continued*)
Merits of Piping Materials of DC Piping Networks

No.	Type	Advantages	Disadvantages	Standards
5	PVC	• Lightweight • Low cost • Available in sizes up to 1200 mm (48 in.)	• Low strength • Poor flexibility • Loses strength at elevated temperatures • Brittle at low temperatures • Cement/field lay up or mechanical joints • Cemented joint must be clean and dry • Difficult to add branch • Water hammer can crack piping • Requires thrust blocks • Low velocity limits	• AWWA C900 (2007b) • AWWA C905 (2010) • ASTM D1785 (2012c) • ASTM D2241 (2009b)
6	Copper	• Good flexibility • Joining by soldering for high integrity joint • Corrosion-resistant but may require protection • Familiar material to workforces	• Expensive • Available only in small diameters: 150 mm (6 in.) and smaller	• ASTM B88 (2009a)
7	HDPE and PE	• Lightweight • Very flexible • Fusion welded for high integrity • Available up to 1600 mm (63 in.) • Leak-free and fully restrained (no anchor blocks)	• Cost depends on petroleum prices • Low strength compared to steel and GRP/GRE • Needs significant wall thickness so increased cost • Reduced inside diameter • Higher pressure losses so may require larger sizes for same flow rates • Larger diameter fusion-welding machines have limited availability	• AWWA C901 (2008b) • AWWA C906 (2007c)

Source: *District Cooling Guide*, chapter 4, 2013. ©ASHRAE www.ashrae.org.

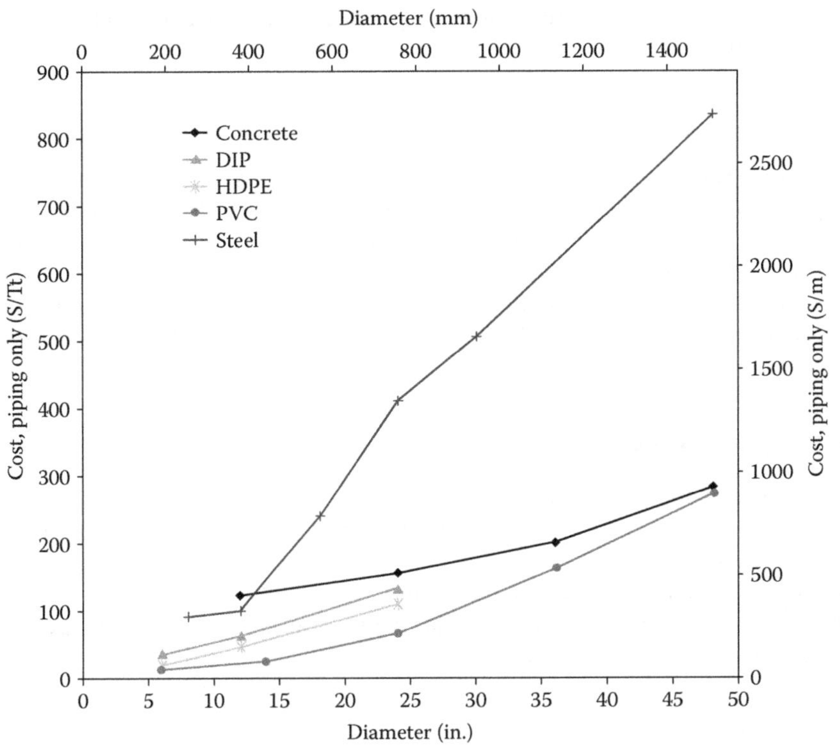

FIGURE 5.10 Relative cost of piping, uninsulated. *Note:* Includes joining. Not included: design, supervision, fitting, excavation, backfill, and surface restoration. (With kind permission from ©ASHRAE www.ashrae.org, *District Cooling Guide,* 2013, chapter 4.)

5.2.1 Carbon Steel and Ductile Iron Pipes

Carbon steel and ductile iron pipes are highly susceptible to corrosion. The external corrosion is dealt with by good preparation of the surface area, painting with anticorrosion paint prior to thermal insulation and water proofing. Chilled water will slowly corrode the inside of steel and iron piping, although chemical treatment of chilled water will delay corrosion. Preinsulation of chilled water piping is made by injecting foam or by spray insulation techniques prior to wrapping by a PE (Poly Ethylene) layer to protect the insulation and act as a water barrier. Cathodic protection is required when the pipes are directly buried. The major advantage of steel and iron pipes is the ready availability of skilled workforce for erection. Good jointing is also available by welding, which can guarantee a solid joint. It is flexible, and its high tensile strength is an added advantage. In Figure 5.10, a clear disadvantage of steel piping is the cost of larger sizes. Large increases in cost occur with steel pipes at large diameters compared to other types of piping material. This and the slow installation associated with steel pipes will generally increase the cost dramatically for larger diameters over 380 mm (15 in.). Ductile iron piping is joined by mechanical systems only, which increases the cost of fittings. Although its moderate strength, flexibility, and leakage rates are within standard limits, these are still disadvantages.

5.2.2 Glass-Reinforced Plastic and Epoxy

GRP and GRE piping are light in weight and possess a high strength, which makes them easier to install and saves on installation time and cost. Their disadvantages are their cement joining in situ or by mechanical means. Adding a branch is also not easily achieved, because its slow curing at low ambient conditions. Linear expansion of GRP and GRE pipes is larger than steel pipes, which creates stresses in the network. It is available in all sizes. Specialized workforce requirement makes it more difficult to find easily.

5.2.3 Cement Pipes

Cement pipes on the other hand are familiar to workforces. They possess reasonable strength and are available in all sizes. The disadvantages are their heavy weight and poor flexibility.

In addition, they are joined by mechanical joints only with difficulty in adding branches. There are limits on their operating pressures and flow velocity. Leakages are allowable according to standards.

5.2.4 PVC Pipes

PVC pipes are low cost, lightweight, and available in sizes up to 1200 mm (48 in.). Their main disadvantages are their brittleness at low temperatures, poor flexibility, and low strength. In addition, they are joined by cement or by mechanical joints with cleanliness an important factor that is difficult to achieve in the field.

It is difficult to add a branch, and its low strength makes it possible to fracture when water hammer occurs. Limits on flow and velocity are recommended.

5.2.5 Copper Pipes

Copper tubes are flexible and joined with high integrity by soldering. They are also corrosion-resistant with some corrosion protection and are familiar to workforces. Their two main disadvantages are their high cost and availability in small diameters, only up to 150 mm (6 in.).

5.2.6 HDPE and PE Pipes

Both HDPE and PE pipes are lightweight, very flexible, and easily fusion welded with high integrity, available up to 1600 mm (63 in.) and are free of leakages, and no anchor blocks are required. However, their cost depends on petroleum prices, which can vary a great deal. Compared to steel, its strength is low and needs a large wall thickness, which increases cost. For large sizes, fusion-welding machines have limited availability.

Higher pressure losses may be expected for larger sizes because of an increased wall thickness at the same flow rates.

When selecting a certain type of piping material, we have to consider all the above in order to select the most suitable type for a specific project. The availability of all these factors must be taken into consideration and which of *these* may affect most the project economics and execution timetable. In some instances, a certain type of material may be the most economical and fit the required project time frame but is discarded because a skilled workforce or other important aspects are not available. In addition, type of piping distribution system used may dictate the type of material used.

A well-protected tunnel-routing system may allow for the use of a piping material that is not easily jointed, because its inspection is easily done. It may also allow the use of a material that does not support compressive stresses; which cannot be used in a directly buried routing. Each project has to be examined separately, and the type of material has to be chosen according to the strength and weaknesses of the routing system, availability of material, economics, and the type of piping material available.

5.3 TYPES OF DISTRIBUTION PIPING SYSTEMS

Many piping distribution systems exist. One of these is the overground piping system. Piping is routed and protected from ambient conditions. This is done by installing an outside jacket made of steel sheeting or aluminum. Water barriers on the thermal insulations is an important addition underneath the protective metal covering. Stainless steel and fabric impregnated with cementing agents have been used. A steel structure usually carries the pipes and helps stacking more than one set of pipes over each other.

This is the cheapest system of distribution available, although it suffers from a low aesthetic view and susceptibility to damage and is seldom used, and other systems are usually preferred.

5.3.1 Directly Buried Preinsulated Pipe Systems

Directly buried preinsulated pipe system are one of the most economical systems when the district cooling network is long and has a large refrigeration capacity. Compared to a tunnel system, it is cost-effective if there are no provisions for a utility tunnel at the design stage. It is also possible to use this system when there are many other utility conduits in the ground to avoid interference, as a deeply buried system is then needed. Detailed site surveys are required to route the system with the minimum of interference. It is important to avoid getting too close to hot piping in existing utilities to avoid heat gains. Figure 5.11 shows a cross section in a directly buried routing for a four-pipes district cooling system.

It is important to note that this system requires good soil investigations to know the type of excavation, and backfilling work is involved. Compaction of the backfill is important as well as providing a drain at the bottom of the trench, especially when the trench is located underneath a street or in an area where irrigation water or heavy rain may cause the compaction to be affected, and soil sagging may occur.

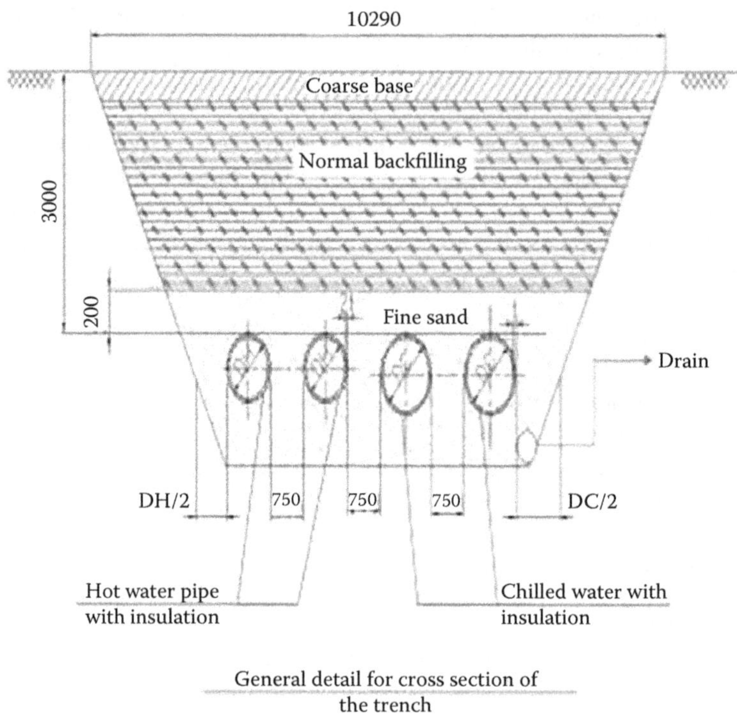

FIGURE 5.11 Directly buried distribution piping system for a four-pipe system.

5.3.2 Accessible Concrete Trench Systems

This system is cheaper to construct than a directly buried system. However, it needs an easily accessible route to allow for the removal of the trench cover when needed. Controlling leaks is possible and can be fixed from the ground level, which allows for fast repairs. It is also not as attractive aesthetically compared to a buried system, a tunnel system, or a deep buried trench system. It is most suitable for a small capacity district cooling system with a few branches and a simple route that allows routing the trench in areas without heavy traffic, therefore protecting the trench from damage and allowing for easy access (Figure 5.12).

5.3.3 Deep Buried Trench Systems

When it is not possible to use an accessible concrete trench system because of heavy traffic in the district or objections to its aesthetic view, a deep buried trench system can be used for district cooling capacities that are not too large. This system protects the piping especially if the piping material does not possess qualities to withstand compressive stresses or high tensile strength.

In these cases, a deep buried trench becomes an attractive solution. Although the system, has its disadvantages since accessing a certain pipe in a certain location to

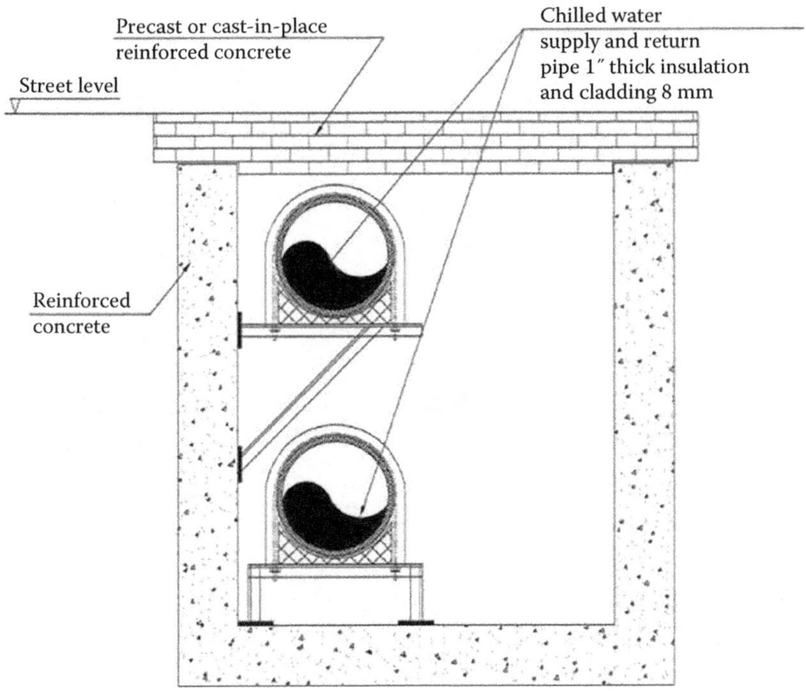

FIGURE 5.12 Accessible concrete trench distribution system.

perform repairs or to replace pipes means excavating, removing the soil, and replacing and backfilling it.

It is nevertheless a good system to use with both small and medium capacity pipes when they need to be protected from surrounding threats. Figure 5.13 shows a cross section of a deep buried trench system.

5.3.4 Drive-Through or Walk-Through Tunnel Systems

This is probably the most economical system and the easiest to maintain the utilities placed in it when it is included at the design stage. The tunnel is constructed underground and allows technical personnel to walk through, or in a long tunnel to drive through. The tunnel allows easy inspection of the utility piping and cables suspended in it or hanging from the walls. It is important to allow a drainage system to evacuate washing water and ground seepage water.

Although the system has the highest capital cost compared to all the other systems, it pays back with the lowest operating and maintenance costs over the lifetime of a project. Tunnel systems are monitored by CCTV or visually by maintenance crews. Figure 5.14 shows a typical tunnel cross section, and Figure 5.15 shows a photo of a tunnel.

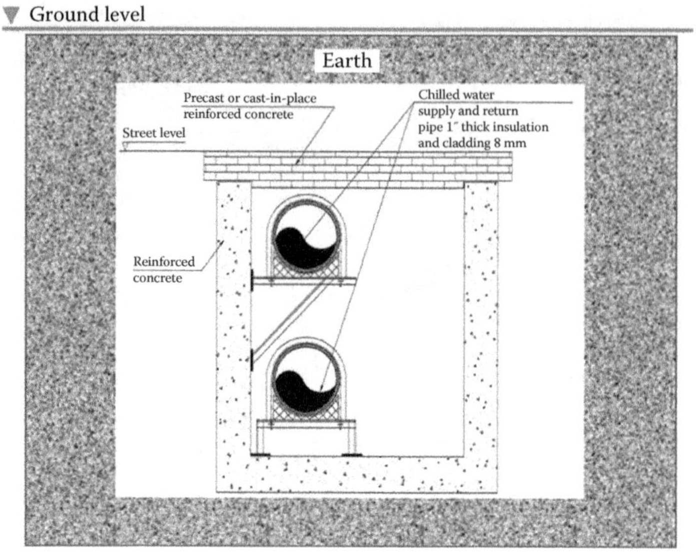

FIGURE 5.13 Deep buried trench distribution system.

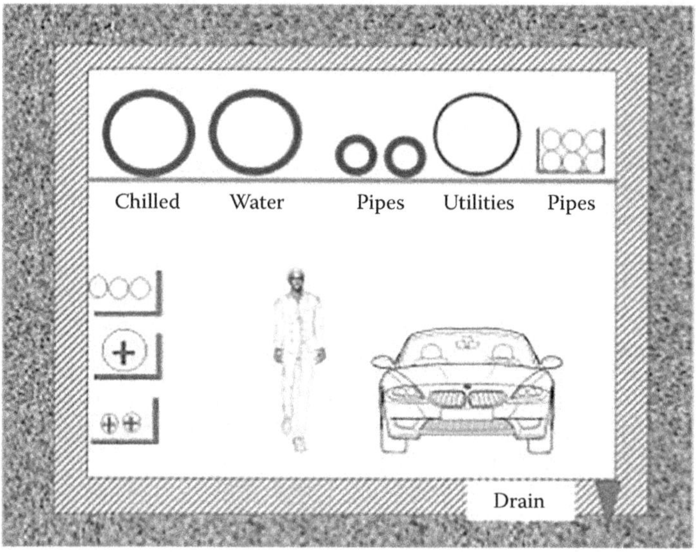

FIGURE 5.14 A walk- or drive-through trench distribution system.

FIGURE 5.15 A drive-through tunnel.

5.4 CATHODIC PROTECTION AND LEAK DETECTION

Steel pipes, directly buried in soil, corrode. The chemical–electric reaction that happens when the soil becomes wet resembles the action of a dry-cell battery where current flows from anodes to cathodes in an electrolyte formed by the wet soil. The anode area where the current is generated destroys the material, and corrosion slowly creates holes in the pipe. This also occurs for preinsulated pipes directly buried in soil at entrances to buildings' mechanical rooms and at connections to valve chambers when the pipe insulation is damaged or repairs are badly applied. It is therefore important in those cases to provide cathodic protection by sacrificial anode systems or impressed current systems.

5.4.1 Cathodic Protection by Sacrificial Anode Systems

When a piece of zinc, magnesium, or aluminum is fixed to the surface of the steel pipe, it forces the steel to become the cathode and the sacrificial metal the anode. This artificial anode becomes corroded as current is generated from it, thus protecting the steel pipe. It is important to use isolation dielectric gaskets at the piping connection that is not preinsulated and is protected. This is because other connected steel that is not isolated will be part of the system, and will weaken the cathodic protection, which makes it useless. Care must be taken to check and replace the sacrificial anodes when necessary.

5.4.2 Impressed Current Systems

In these systems, a current is applied to the steel surface. Inert anodes are fixed to the surface of the steel pipe to create an artificial anode with the impressed current, thus protecting the steel surface. A rectifier is used to convert the alternate current of the source into the direct current, and the current intensity is periodically adjusted as the steel pipe ages and holds less electrons.

5.4.3 Leak Detection Measures

Several leak detection methods are used for chilled water systems. The simplest ones are copper wires, which may be insulated or uninsulated depending on the design. They are placed inside the foamed insulation of the pipes. The detection relies on looking for a short-circuiting caused by the water surrounding the wire and monitored by electric resistance or impedance change. The DC building management system (BMS) monitors leaks continuously, and when a leak occurs, it is reported, located, and fixed. The wires run along the entire pipe's length and connect at junctions.

6 Designing Energy Transfer Stations

6.1 TYPES OF CONNECTIONS TO END USERS

District cooling systems are connected to distribution networks. These in turn are connected to end users by one of the two methods:

- Direct connections
- Indirect connections

6.1.1 Direct Connections

Both types of connections are used successfully. The type of connection used depends on the nature and application of the district cooling system.

In a direct connection, the same chilled water produced circulates in the DC plant and the distribution network. Therefore, there is no interface between the chilled water of the plant and the in-building distribution network, and hence no separation of chilled water between the production, distribution, and in-building HVAC system.

6.1.2 Indirect Connections

In an indirect connection, an interface is used, usually a plate heat exchanger. Plate heat exchangers are the preferred heat exchangers in DC systems because traditional shell and tube or shell and coil heat exchangers are bulkier when they are designed to operate at the small approach temperatures in use in DC systems. These are normally 0.5 to 2°C. In addition, traditional heat exchangers are often more costly. Space limitation is also an issue. Space is limited in DC buildings' mechanical rooms and is at a premium, especially in commercial and administrative applications. Rent is often considerable.

Both direct and indirect connection systems are applicable in DC systems. Table 6.1 compares both connection types and shows the important features of each type. Those are summarized in the following 10 features:

1. Achieving high ΔTs in chilled water (CHW) distribution networks
2. The limitation of using vapor compression chillers compared to absorption chillers
3. CHW quality control
4. The battery limit responsibilities between the service provider and end users
5. Cost implication of both connections
6. Retrofitting a DC system to an existing district

7. The impact of the type of connection on building CHW pressure levels
8. The differentiation between a DC system where the provider and end user are the same and that with other ownership schemes (see Section 2.1)
9. The space requirement for the connection
10. The need to install pumps or not inside buildings near the DC station

Table 6.1 summarizes these 10 features and differentiates them. It is important for the designer to judge which of these features are more important and applicable for a certain district and then choose the most suitable type of connection for the DC system.

TABLE 6.1
Comparison between Direct and Indirect End-User Interfaces

No.	Features	Direct Connection	Indirect Connection
1.	Achieving high ΔTs in circuit distribution	No HX approach temperature needs. Greater ΔTs in building, hence easier control of humidity	HX needs 1–2°C approach, reduces ΔTs in building
2.	Use of vapor compression or absorption chillers	Absorption chiller operates at or near 5°C—limits ΔT of plant	Larger ΔT of plant possible, vapor compression chiller operates at lower temperatures
3.	Chilled water quality control	No isolation between CHW production, distribution, and end user; quality control must be maintained rigorously	Isolation between CHW production, distribution, and end user; quality control easier to maintain
4.	Division of responsibilities	The battery limits contractually, between provider and end users are not clear. May cause conflict when disputes arise	Battery limits are clearer contractually. Helps restrict disputes between provider and users
5.	Cost	Lower capital cost	Higher capital cost
6.	Retrofitting existing system	Easier to use in applications where DC is introduced to an existing development	Harder to use in retrofit applications
7.	Independent pressure at buildings	Isolation is achieved by applying pressure, isolation techniques for tall buildings, and high DC operational pressure differentials	Isolation of pressure exists between DC system and buildings
8.	One DC user	May be preferable for one user/customer applications	Can be used for one customer but becomes more expensive
9.	Space requirements for in-building mechanical rooms	Needs little space for in-building mechanical rooms	Needs larger space for in-building mechanical rooms
10.	Need for in-building pumps at adjacent buildings	Building near DC station may not need a CHW pump	All buildings will need CHW pumps

Many names are attributed to end-user connections such as an end-user interface or consumer interface. The name that is most in use is, however, energy transfer station (ETS).

ETS includes, besides flow control valves, measuring instruments such as flow meter pressure sensors, temperature sensors, a means to enable calculation of energy consumed in the building, and energy meters.

6.2 OPERATION OF DIRECT CONNECTIONS

Figure 6.1 shows a simplified diagram for a chilled water direct connection type for a building. The figure shows a piping connection between the return and supply, regulated by a two-way valve on the return pipe. Flow in this connection allows the return water to recycle or return to the network in proportional settings, thus controlling the supply chilled water flow to loads. Regulating piping also decouples pressure between the building and the distribution network.

Two-way valve control loads and variable speed pumps are used to control flow rates. During periods of low loads, the flow rate is controlled by a bypass connection at the end of an in-building piping system.

This connection system is quite common and operates especially well when a low ΔT syndrome is well controlled (see Chapter 3).

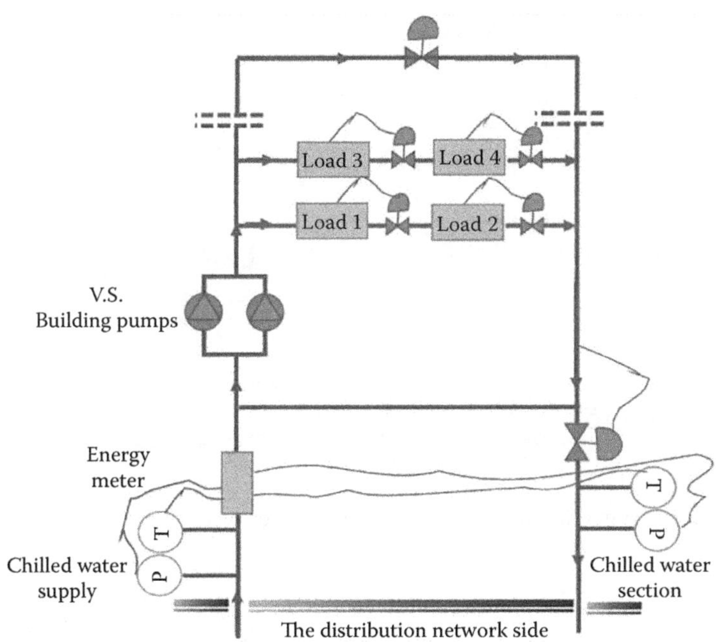

FIGURE 6.1 A simplified diagram for a direct connection of an in-building chilled water system.

6.3 OPERATION OF INDIRECT CONNECTIONS

Figure 6.2 shows a simplified diagram of an indirect chilled water connection. In this figure, two plate heat exchangers are used. The number of plate heat exchangers used is defined by the need for redundancy and the total refrigeration capacity of the building. This arrangement allows for the total separation between the distribution chilled water network and the building chilled water. Large systems, over 8800 kW (2300 TR), use more than one heat exchanger.

In the figure, two control valves are used to control temperature on the customer side or "hot side" (see the sensor on each valve connected to the supply pipe of heat exchanger on the hot side). One on each return is placed on the "cold side" or utility side of the heat exchanger.

The variable speed pumps (VS) are controlled by the two-way valves at the loads. A minimum flow rate control valve is used at the end of the building chilled water circuit to provide flow during low-load conditions. This arrangement is quite common. Its selection will depend on the district conditions as shown in Table 6.1.

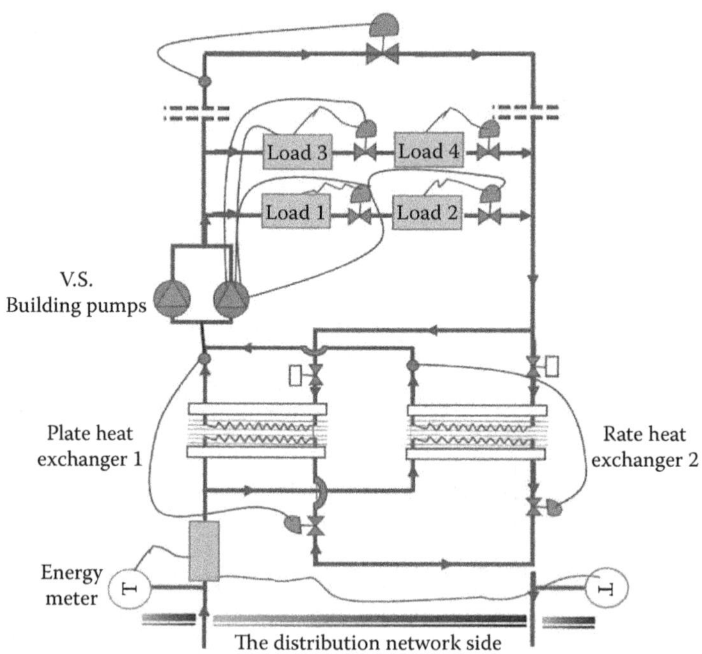

FIGURE 6.2 A simplified diagram for an indirect connection of an in-building chilled water system.

6.4 METERING AND ENERGY METERS

To measure energy used by end users, energy meters are installed in the building's mechanical rooms. Energy meters utilize equipment for measuring flow, temperature differences between supply and return of chilled water, and the time duration between two readings using an energy calculator.

There are two types of energy meters: dynamic and static.

6.4.1 Dynamic Energy Meters

6.4.1.1 Impeller Meters

These meters use an impeller fitted with blades. These energy meters are sensitive to sand and sediments that might be entrained with chilled water. They are suitable for medium-sized buildings because they do not operate well at small loads.

6.4.1.2 Turbine Meters

Turbine meters use a rotor to measure flow. They need a sizable flow to operate and are therefore more suitable for larger capacity buildings. Accuracy ranges from 1.5% to 3%.

6.4.2 Static Flow Meters (MID Meters)

6.4.2.1 Magnetic Induction Meters

These meters induce current in the water using electrodes connected to electromagnets. The flow is measured by measuring the voltage. This is converted to flow rates.

As the measuring method is static, less maintenance is needed when compared to dynamic meters. Initial cost is, however, higher than dynamic meters.

MID meters are successfully used in district cooling applications. Their accuracy is high: ±1% to 2%.

6.4.2.2 Ultrasonic Meters

Ultrasonic meters use the changes in the propagation of ultrasonic waves to measure flow. Ultrasonic meters proved accurate and suitable for large flow rates.

Figure 6.3 shows a schematic diagram of an indirectly connected system with components of an energy station. Figure 6.4 shows a schematic diagram of a directly connected system with an energy station. Figure 6.5 shows an expanded view of a plate heat exchanger. Figure 6.6 shows an energy transfer station with HX and pumps.

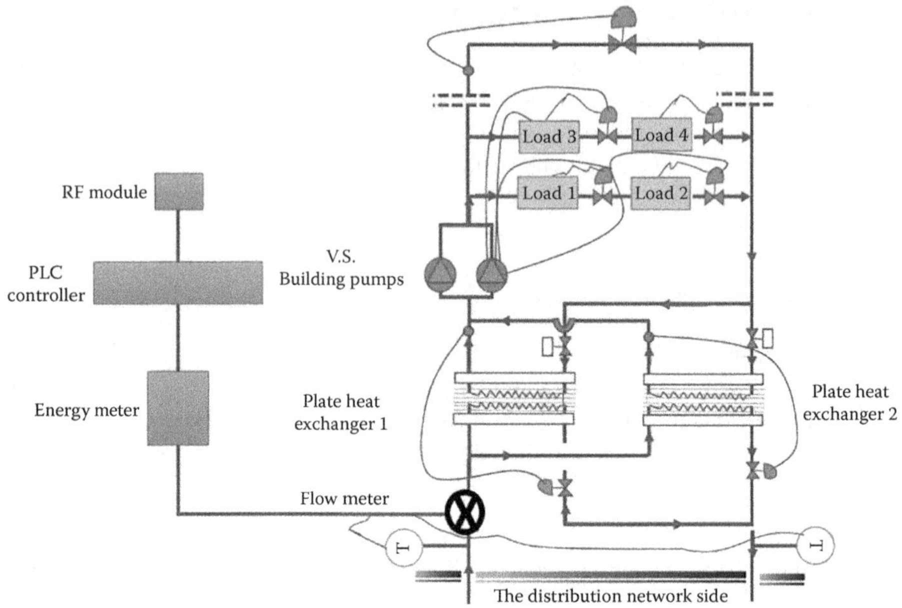

FIGURE 6.3 In-building chilled water indirect connection with an energy meter for drive-by meter reading.

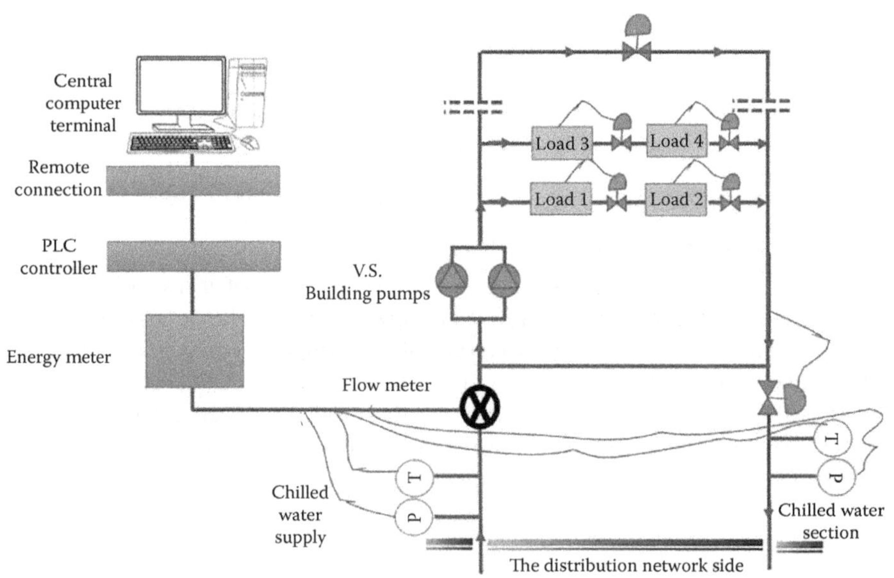

FIGURE 6.4 In-building chilled water direct connection diagram with an energy meter for a remote connection.

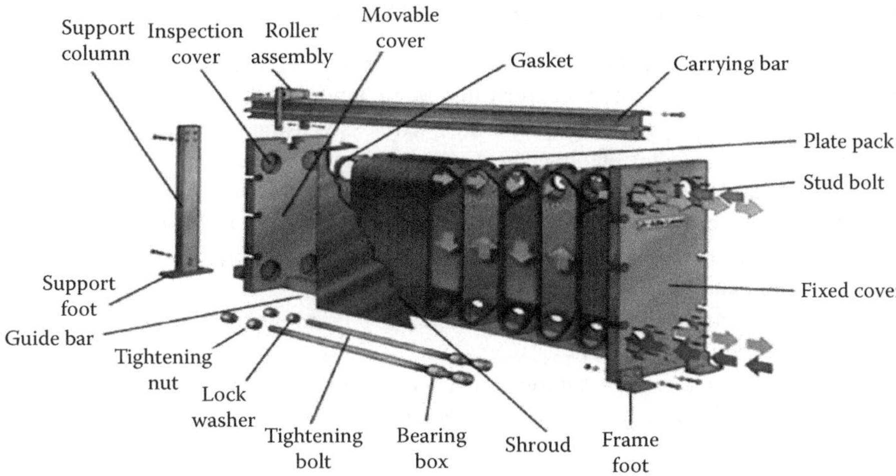

FIGURE 6.5 An expanded view of a plate heat exchanger. (Courtesy of Alfa Laval.)

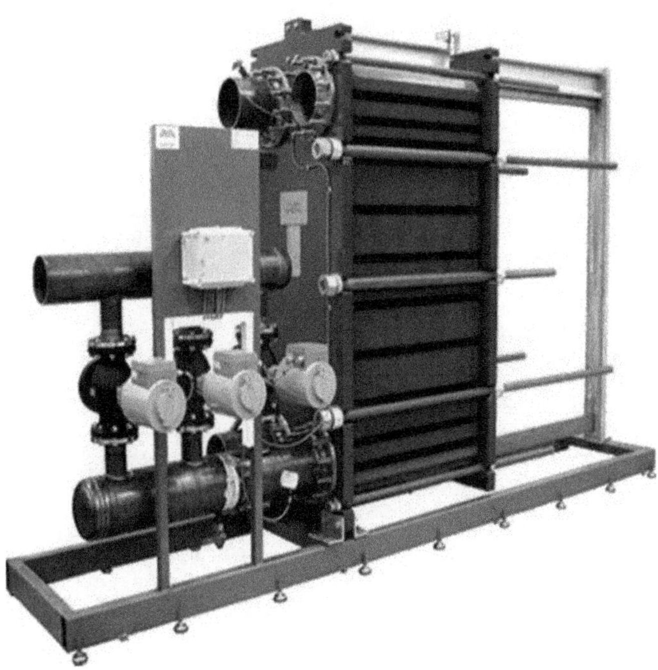

FIGURE 6.6 An energy transfer station with HX and valves. (Courtesy of Alfa Laval.)

6.5 COLLECTION OF DC METER READINGS

Energy meter data are collected either at the meter or at a remote location. Local readings of meters use a handheld terminal that connects to the meter.

Remote energy meter reading operates wirelessly using a radio signal from a device in the meter, via the telephone network, or via an Internet connection. Figures 6.3 and 6.4 show some of these connections.

In energy meters fitted with radio frequency modules, a radio frequency concentrator connected to a central computer uploads the data, and bills can be produced for each end user.

In meters connected via the Internet, meters are fitted with a TCP/IP module and are read by a central computer.

Often there is a need for submetering, when a building is rented to more than one end user. In this case, a secondary submeter is needed or water meters are used at end users to measure flow rates and allocate submeter reading proportionally according to water flow meter readings. This method is more economical than using submeters and is cost-effective.

Another method used by some DC providers involves calculating individual consumption by floor area of the space instead of submetering. This method does not provide an incentive for end users to conserve energy.

7 Design of Thermal Energy Storage

7.1 DEFINITION OF TES

Thermal energy storage (TES) stores cooling enthalpy during off-peak times to use during on-peak times. A specially constructed insulated tank stores the cooling energy at off-peak times and uses it at on-peak times.

This technique allows for the use of fewer chillers during on-peak times than those necessary to cope with peaks on the daily cooling load demand curve (see Section 3.3).

The rating of TES is based on its ability to hold a certain refrigeration capacity for so many hours. For example, a 20,000 TR.h capacity TES will hold 10,000 TR for 2 h or 5,000 TR for 4 h or other combinations totaling 20,000 TR.h.

TES systems have been successfully incorporated in District cooling systems for many years. TES is accepted as an integral part of a DC system.

Applications range from universities, colleges, airports, museums, sport complexes, and hospitals to leisure centers and administrative buildings; military facilities use TES as do many other applications.

7.2 BENEFITS OF TES

There are many benefits of installing a TES system in a district cooling application; are discussed in the following sections.

7.2.1 Shifting On-Peak Cooling Load Demand to Off-Peak

This is probably the most important benefit of a TES system.

Figure 3.3 shows a typical daily cooling load demand profile and the average daily load and charging/discharging loads. The figure shows that the charging of the TES is done at 11.30 pm to 9.30 am the next day, which satisfies the district load for the day including on-peak periods with chillers continuously operating at the average load. This reduces electric demand charges, thus reducing operating costs. Cash incentive payments by the local utility could also provide a payback for the reduction of on-peak electric demand.

7.2.2 Reducing Installed Cooling Capacity

Figure 7.1 shows that the average capacity is 73% of the total installed capacity of a district cooling plant, according to the typical daily cooling demand profile. Including for a 27% capacity TES designed for 10 h operation will reduce the capacity of

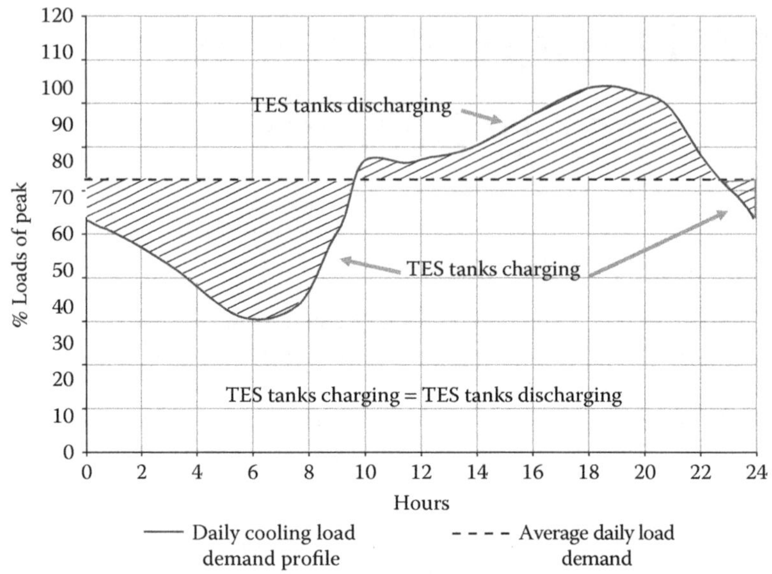

FIGURE 7.1 Typical daily cooling load demand profiles showing average daily load and charging/discharging loads.

installed chillers by 27%, excluding standby chillers in both cases. Those will have to be always included in the plant to guard against a chiller malfunctioning or programmed maintenance periods.

Although the cost of a TES system varies a great deal from country to country and from time to time, allowing for a TES system is usually beneficial for the capital cost investment. Discarding the cost of additional extra chillers and ancillary equipment will usually yield a savings in capital costs.

7.2.3 Improving Plant Economic Performance

A TES charging cycle operates at off peaks, which are usually late night and early morning hours. These hours during the night are when peak ambient temperatures are at their lowest compared with on-peak ambient temperatures when they are much higher especially in high-ambient temperature countries. Operating chillers at those off-peak temperatures results in better economics due to better chiller efficiencies. This reflects on the overall plant power consumption, improving its overall efficiency and reducing power consumption. It is also important to note that the almost flat capacity percentage of the load allows the efficient use of a cogeneration system, producing both electric power and cooling capacity, when those systems are economically viable. This is because cogeneration systems work better with a constant supply of electrical power and the corresponding heat for use in cooling and heating applications.

7.2.4 TES Tank Use as an Emergency Chilled Water Source

A TES tank can be utilized as a redundant chilled water source in emergencies. This is possible in applications where chillers breakdown, and the resultant loss of cooling capacity is not tolerated, which is often the case in DC schemes.

7.2.5 Other Special Uses of TES

A TES tank has been used as a standby fire protection reservoir in some DC applications, by connecting it to the fire sprinkler networks, or other means of fire extinguishing.

When a TES system, which is normally positioned near the DC plant, is positioned in remote locations, the TES tank will act as a second DC plant. This so-called second DC plant will switch online on the distribution network at times when the original plant cannot cope with cooling loads at remote buildings or for building loads added at the last minute and not anticipated originally.

In some special applications, it becomes necessary to increase the DC system's ΔT of the distribution network. This may be because of added loads on the system or because the piping network was undersized at the design stage. Adding a TES tank together with a plant capable of operating at a larger ΔT than the original DC system's ΔT and then mixing both systems' chilled water will increase the ΔT of the combined system, thus increasing the cooling load.

This is also possible when a system is redesigned to operate at "low-temperature air" to reduce the size of its air handlers, fan coil units, ductwork systems, and pumps. Cold air systems at end-user buildings allow the selection of smaller air handlers, fan coil units, and distribution ductwork. This in turn reduces capital costs.

This is often the case when retrofitting an older HVAC system to transform into a DC system, and the existing "air side" cannot cope with increased loads. The lower ΔT created by using a dedicated TES tank will help achieve an overall larger ΔT for the system.

7.3 TYPES OF TES

There are three major types of TES systems: chilled water, ice and slurry, and low-temperature liquid storage systems. The three systems are explained below, and their advantages and disadvantages are also shown.

7.3.1 Chilled Water Storage Tanks

This is the most common type of TES. This system stores chilled water, making use of its sensible heat. Its capital cost is considerably less than ice storage systems of equivalent capacity.

Chilled water storage tanks are easy to operate, and their control system is not complicated, with chilled water stored at 4°C–6°C. The most common type of chilled water storage tanks are the stratified type. In this system, in the discharge cycle, the tank is filled at its top with warmer water, whereas colder water is taken from its bottom. Because warm water has a lighter density than cold water, stratification occurs between the two masses and the masses remain separated.

Water is filled into the tank using special nozzles distributed inside the tank to preserve stratification by reducing water speeds.

The main disadvantage of a chilled water TES tank is its large size. Although TES-chilled water tanks can be designed in concrete or steel and isolated thermally, both their sizes are bulky. In locations where land prices are high, this may constitute a service drawback.

However, they are the most common type of TES systems. They are available in capacities on an average of 70,000 kW.h (20,000 TR.h) and exist in capacities of up to 2,000,000 kW.h (600,000 TR.h).

7.3.2 Ice and Slurry Storage Tanks

The second most common TES system is ice storage. In this system, the latent heat of water reduces the size of the tank, thus saving valuable land area. The technology is well mature and allows storage of chilled water at 1°C to 4°C when ice is mixed with chilled water.

The system is, however, more expensive than chilled water storage. The system's important advantage is that it allows the DC system's ΔT to be increased, thus reducing the capital cost of the distribution network and ETS.

Ice-producing chillers are costlier than ordinary chillers operating at higher chilled water temperatures. This may affect the economic considerations of the system.

Ice slurry, a mixture of ice and chilled water, is used in TES. The mixture is possible to pump. The system's uses are still limited, although it has an advantage of having a lower storage temperature than ice-only systems.

7.3.3 Low-Temperature Liquid Storage Tanks

In this system, special chemical additives are added to chilled water to allow storage at temperatures down to −1°C to 2°C. This allows the use of smaller storage tanks when compared to normal chilled water storage, thus saving on the capital cost.

The system is suitable for increasing chilled water ΔT. It is more capital intensive when compared to normal chilled water TES, because it needs special low-temperature chillers and heat exchangers between the TES tank and the distribution network for keeping the low-temperature liquid in the TES storage circuit.

7.4 DESIGNING TES SYSTEMS

Designing TES systems depends on the philosophy of operation that will be adopted to operate the system.

This could be one or more of the following cases:

1. On-peak shaving of cooling capacity, as is the case when the electric demand charges are high, and there is a need to shift the on-peak cooling load to off-peak periods
2. Use of TES as an emergency reserve when not enough cooling capacity is achieved

3. To operate at times when the daily cooling load demand curve is different from those of the usual daily pattern as in weekends or annual vacations
4. To operate as a backup to the plant chillers to compensate for a chiller breakdown
5. To use in a remote position from the DC plant to help cope with cooling loads added or not anticipated when the system was originally designed

It is possible to design a TES system to operate during periods of the day when the chillers are totally shut off. This kind of full-duty TES can be very important if it is necessary to reduce electrical power greatly at certain peak times in the day. In this case, the chiller's installed capacity will have to be much larger than average capacities to provide enough capacity to charge the TES tank(s) during the daily chiller's total shutdown period.

However, TES tanks that operate to compensate for partial duty during on-peak periods are much more common.

TES tanks are designed to absorb heat gains of less than 1% of their cooling capacity per day. It must be noted that the inclusion of a TES tank to an existing DC system will require an additional cost related to the additional control equipment necessary to allow the automatic operation of the TES tank with the DC system.

8 Controls and Instrumentation

8.1 INTEGRATED CONTROL AND MONITORING SYSTEMS

An integrated control and monitoring system (ICMS) is necessary to control and monitor the various components of a DC plant, distribution system, and ETSs at various buildings. The industry standard is a supervisory control and data acquisition (SCADA) system, with PLC microprocessor-based controllers.

A building management system (BMS) can also be used with microprocessor-based controller (DDC) that has somewhat a lower performance but is more economical than a SCADA system.

There are four layers of ICMS:

1. Management layer
2. Communication layer
3. Automation layer
4. Field instrument layer

Each of these layers performs vital functions and together they provide total control and monitoring requirements.

The *management layer* has a central operator workstation(s) and will allow accessing parameters of various parts of connected equipment. It is able to monitor the ETS controllers, fire alarm system, and BMS at the DC plant building. All events and values can be stored for a later review and analysis.

The *communication layer* is an Ethernet communication to connect controllers with management by a fiber-optic cable. The operator can read ETS events and values as they happen and energy measurements can be known. RF modules can also obtain reading of ETS; on a drive-by basis (see Chapter 6) at this layer, all interfaces to the main equipment and other systems are provided.

The *automation layer* comprises microprocessor PLC controllers with ancillaries and a redundant stand-alone capability, when connection to management level fails.

The *field instrument layer* includes all industrial grade field instruments including flow meters, actuators, and temperature transducers.

The system monitors or controls the following:

- Chiller plant parameters
- Cooling tower plant parameters
- TES(s) tank parameters
- Chemical treatment equipment parameters

- BMS at DC plant building
- Firefighting status at DC plant building

Other functions may be added such as

- Pumps status
- Valves status of distribution network
- Makeup water for cooling tower evaporation and blowdown
- Status of filtration of chilled water and condenser water
- Expansion tanks and air separators status
- Utility power distribution status

8.2 THE CONTROL STRATEGIES FOR DC PLANT EQUIPMENT

This is an example of control strategies for a safe and efficient operation of DC plant equipment.

8.2.1 Chilled Water Supply and Return Temperature

Continuous monitoring/control of supply and return chilled water temperatures to help achieve design parameters and avoiding low ΔT syndrome.

8.2.2 Chiller Monitoring and Control

Control operation of chillers at or near their best operational efficiency. Allow stability of chilled water temperature and allow cycling and sequencing in a safe and efficient way.

8.2.3 Cooling Towers: Monitoring and Operation

It is important to control the operation of cooling towers at their best operational efficiency. Sequence and cycle control are necessary for their safe and efficient operation.

Operators should be provided with viewing trends and the flexibility to change set points. This ensures that the plant abides by the limits on electrical demand capacity. Ensuring the sequences of operations are followed, while control loops provide necessary safety interlocks for chillers stoppage, stopping primary pumps, condenser pumps, and valves. Provide warning alarms when limits of operation are reached.

Control operation of water makeup for the condenser water blowdown and evaporation. Monitor energy flow meters by communicating with energy meters and displaying flow rates. Measure energy consumption at the building level by ETS.

ICMS should provide necessary control strategies in order to ensure that the DC plant operates at its most efficient mode and at its safest conditions.

8.3 OPERATIONAL SEQUENCE OF A DC PLANT: PLANT DESCRIPTION AND DISTRIBUTION PUMPING SCHEME

8.3.1 Practical Example

Assume a DC plant is equipped with two chillers and is divided into two modules. Each module is connected to its primary chilled water pump. One cooling tower is used per module, each is connected to one condenser pump. The chilled water distribution network is a primary-distributed secondary pumping arrangement.

The building pumps, condenser pumps, and cooling tower fans are all equipped with speed controls using VFDs. The connection interface used is a direct connection. Each building is equipped with an ETS incorporating energy meters and controls.

The DC provider monitors the energy meters and controls the secondary-distributed pumps at buildings. An Ethernet system with a fiber-optic cable is used for communicating between a DC plant and ETSs at buildings. The DC system is equipped with a TES tank that helps shave on-peak loads to reduce electric utility demand charges. The supply chilled water temperature is set at 5.5°C.

8.3.2 Sequence of Operation

It is possible to change the chiller module lead/lag selection by choosing which module has run most on a fortnight basis. When a module is switched on, chilled water isolation valves and corresponding cooling tower isolation valves will open. An adjustable delay is built in before both the chilled water and condenser water pumps operate. The condenser pump speed will ramp up until it reaches 100% in 5 min.

After an internal start-up check is performed on the chiller and it shows no abnormalities, the chiller is switched on. The signal from the chiller's water flow switch will indicate the chilled water flow status.

When the load demand reaches 85% of the chiller's capacity and stabilizes, no action is taken. When the supply chilled water temperature starts rising and exceeds 6.5°C for 10 min, the second chiller module will start in the same sequence as the first module.

When the system load demand diminishes below 50% and the supply chilled water temperature decreases to less than 7.5°C, the second module chiller will be stopped.

After a 3 min delay, the cooling tower isolation valves will close and the chilled water isolation valves will also close. In addition, the second module chilled water pump and condenser water pump will stop working. When the module is operational, each condenser water pump in a chiller module will ramp up and down to modulate its speed. This is done to achieve the set design temperature difference of the condenser water.

The set point of the condenser water temperature is controlled by the cooling tower fan speed, which is modulated to achieve the condenser water temperature set point.

The speed of the distributed in-building chilled water pump is controlled by the load demand inside the building; it increases as more loads are switched on and decreases as loads are switched off, and is monitored at the DC plant.

The pump minimum speed shall also ensure that the return chilled water pressure differential to the DC plant will always be higher than the station net positive suction head (NPSH). This is set and monitored at the plant expansion tanks and is controlled by the DC plant operator.

The control system will always ensure that the chillers operate near their best efficiencies. Chilled water production will charge the TES tank at off-peak and discharge it at the on-peak time. The control system shall initiate the charging and discharging cycles of the TES system. It will also monitor the thermal tank capacity on a continuous basis.

A well-designed and operated control system in a DC installation will ensure the following:

- An efficient operation with minimum safety concerns
- Longer equipment lifetime expectancy
- A verifiable clear system for billing customers according to their chilled water energy consumption
- Smooth operation with a minimum number of operators
- An early warning system of possible breakdown of equipment, thus avoiding a service stoppage

8.3.3 Operation and Maintenance

Good operation and maintenance practices are very important to ensure uninterrupted service. Devising a well thought out preventive maintenance program for both the plant equipment and the distribution system will contribute greatly to achieving this goal. Understanding and executing equipment manufacturer's recommendations for maintenance are a part of reaching this goal.

8.4 INSTRUMENTATION

8.4.1 Field Level

- Sensors, detectors, transmitters, and transducers for various parameters; temperature, pressure, flow, and levels
- Thermometers, pressure gauges, level indicators, and so on
- Control valves and actuators
- VFS

8.4.2 Automation Level

- PLC or DDC controller
- I/O module
- Galvanic isolator
- Local operating unit
- Local control panels and accessories

8.4.3 Communication Level

Network devices, that is, routers, gateways, fiber-optic, and control cables with accessories.

8.4.4 Management Level

The PC operator station, process data manager, energy calculation servers, hot standby servers, and software.

Case Studies

THE SMART VILLAGE

General

The system location: Giza, Egypt. The Desert Road to Alexandria—Toll station Cairo end
A modern IT administrative park
District cooling plant: Two plants
Refrigeration capacity: 42,240 + 17,600 = 60,000 kW (12,000 + 5,000 = 17,000 TR)
Expected final refrigeration capacity: 85,500 kW (24,000 TR)
End-user buildings served: 62 buildings
End-user buildings, after all buildings are completed: 85 buildings
District heating source: Absorption unit—Qt: 6
Capacity of each absorption unit in heating mode: 2800 kW (800 TR)
District heating water temperature, in/out: 55°C/60°C

Chiller Plant Details

Types of chillers: Natural gas-fired absorption chillers and vapor compression centrifugal chillers
Cooling of chillers: Water-cooled
Number of chillers: 12 absorption chillers and 2 centrifugal chillers
Nominal capacity of chillers: Absorption chillers 4200 kW (1000 TR each) and centrifugal chillers 8800 kW (2500 TR each)
Refrigerants: Absorption chillers, water (LiBr-H_2O solution); centrifugal chillers, HFC 134a
Chillers arrangement: Parallel counterflow

District Cooling System

Build, own, and operate: 20 years
Yearly total energy supplied (chilled water), TR × h/y: 22,000,000
Chilled water design ΔT: 7.7°C (14.0°F)
Chilled water supply/return °C: 5.6/13.3

Pumping Details

Number of primary pumps, chilled water (CHW): 14
Capacity of each CHW primary pump: 90 kW
Type of CHW pump: Horizontal split case
Number of secondary pumps, CHW: 9
Capacity of each CHW secondary pump: 200 kW
Type: Horizontal split case
Drive: Variable speed

Tertiary pump's capacity: Depends on building size
Number of condenser pumps: 14
Capacity of each cooling water (CW) pump: 200 kW
Type of CW pump: Horizontal split case

Cooling Tower Details

Position: On top of separate building
Number of cooling towers: 14
Number of cells per cooling tower: 2
Capacity of cooling tower—Absorption unit: 6400 kW (1800 TR)
Capacity of cooling tower—Centrifugal unit: 12,700 kW (3600 TR)
Type: Induced draft—Axial fans
Cooling tower pump—Absorption unit: 1050 m^3/hr
Cooling tower ΔP: 40 m
Cooling tower ΔT: 5.6°C
Cooling tower design CW temperature in/out: 35/29.4°C

Chilled Water Distribution System

Type: Four-pipe system (2 chilled water, 2 hot water)
Directly buried, preinsulated piping system made of carbon steel pipes, insulated, and wrapped in factory by spraying polyurethane foam, and wrapped with polyethylene (PE) layer
Pumping of distribution system: Primary constant flow—Variable secondary flow—Variable tertiary flow
Piping network length (cold and hot): 51,000 m
Size of pipes, at station, main headers: 48 in.
Size of pipes, CHW, at station end: 32 in.
Size of pipes, hot water, at station end: 18 in.
Number of pipe main branches: 4 (A, B, C1, and C2)
Piping material: Carbon steel
Piping and insulation, CHW: Preinsulated, sprayed at factory—1 in. thick with a PE jacket
Piping and insulation, hot water: Preinsulated, sprayed at factory—2 in. thick with a PE jacket
Joining: Electric arc—Welding
Testing of pipes: Liquid plug system and water pressure and X-ray at golden welds
Cathodic protection: N/A
Leak detection: N/A
Number of manholes and maneuvering rooms: 100+

Consumer Interface

Direct connection
Energy meter reading: At thermal energy storage (TES)
Tariff penalty for a low ΔT: N/A

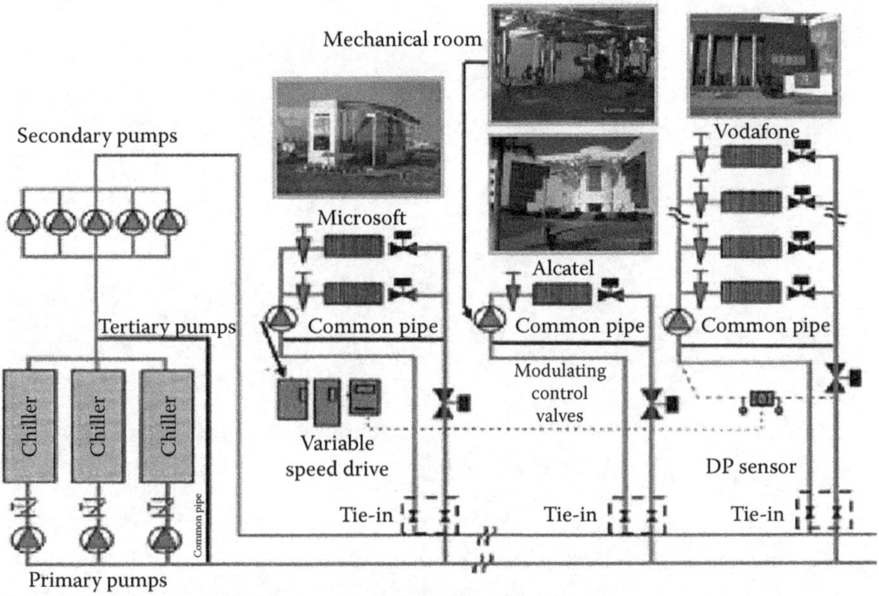

Primary–secondary–tertiary pumping system

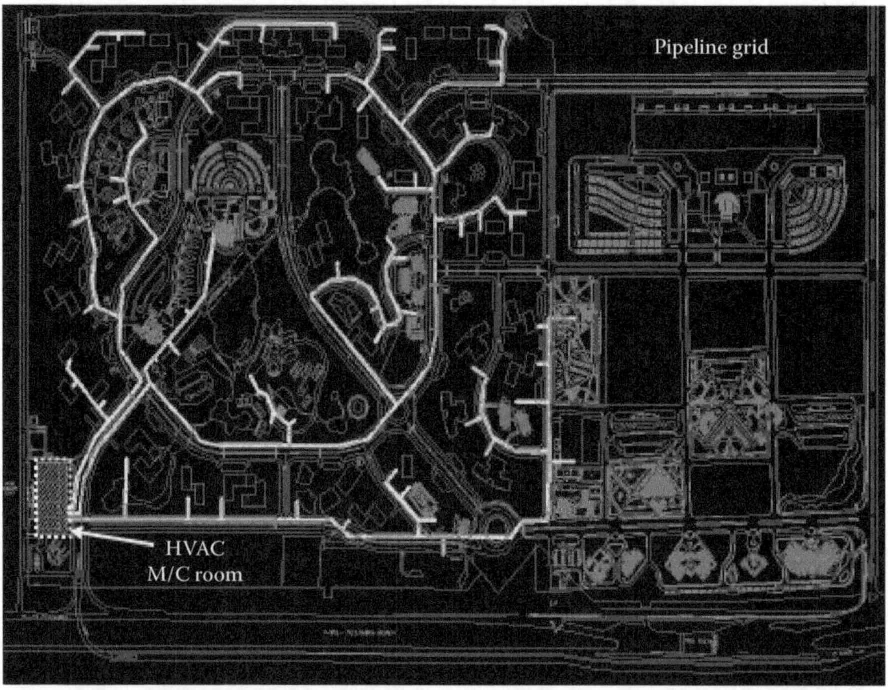
Pipeline grid
HVAC
M/C room

THE AMERICAN UNIVERSITY IN CAIRO

General

The system location: New Cairo, Egypt
All the campus buildings of the university
District cooling plant: One chilled water, one hot water plant connected to a cogeneration plant
Refrigeration capacity: 19,400 kW (5500 TR)
Expected final refrigeration capacity: 28,000 kW (8000 TR)
End-user buildings served: All the university buildings
Buildings that can be added in the future: 5 more buildings
District heating source—Qt: Natural gas-fired hot water boilers, 2
Capacity of each heating boiler: 3600 kW
District heating water temperature, in/out: 55°C/60°C

Chiller Plant Details

Types of chillers: Natural gas-fired absorption chillers
Cooling of chillers: All water-cooled
Number of chillers: 5 absorption chillers
Nominal capacity of chillers: 4 × 4200 kW (1200 TR each) and 1 × 2500 kW (700 TR)
Refrigerants: Water (LiBr-H_2O solution)
Chiller arrangements: Parallel counterflow

District Cooling System

Build, own, and operate: 20 years
Yearly total energy supplied (chilled water) TR × h/year: 6,000,000
Chilled water design ΔT: 7.7°C (14.0°F)
Chilled water supply/return: 5.6°C/13.3°C

Pumping Details

Number of primary pumps, CHW: 5
Capacity of each CHW primary pump: 45 kW
Type of CHW pump: Horizontal split case
Number of secondary pumps, CHW: 5
Capacity of each CHW secondary pump: 90 kW
Type: Horizontal split case
Drive: Variable speed
Tertiary pump capacity: Depends on building size
Number of condenser pumps: 5
Capacity of each CW pump: 160 kW
Type of CW pump: Horizontal split case

Cooling Tower Details

Position: Stand-alone next to DC plant
Number of cooling towers: 5 (4 + 1)
Number of cells per cooling tower: 3
Capacity of cooling tower: 4 × 6400 + 1 × 4600 kW (4 × 1800 + 1 × 1300 TR)
Type: Induced draft—Axial fans
Cooling tower ΔT: 5.6°C
Cooling tower design CW temperature in/out: 35°C/29.4°C

Chilled Water Distribution System

Type: Five-pipe system (2 CHW + 2 HW + 1 service hot water)
Located in a utility tunnel running underneath the university buildings and connecting all the mechanical rooms in the buildings
Pumping of distribution system: Primary constant flow—Variable flow secondary—Variable flow tertiary
Piping network length (cold and hot): 2800 m
Size of pipes, at station, main headers: 24 in.
Size of pipes, CHW, at station ends: 20 in.
Number of pipe main branches: One
Piping material: Carbon steel
Piping and insulation, CHW: Preinsulated, 1 in. thick with PE jacket
Piping and insulation, hot water: Preinsulated, 2 in. thick with PE jacket
Joining: Electric arc—Welding
Testing of pipes: Water pressure test
Cathodic protection: N/A
Leak detection: N/A
Number of manholes and maneuvering rooms: N/A

Consumer Interface

Direct connection
Energy meter reading: At DC plant
Tariff penalty for a low ΔT: N/A

DESIGN OF TURBINE INLET COOLING IN A HIGH-AMBIENT TEMPERATURE COUNTRY

In a gas turbine power station, the performance of gas turbines deteriorate greatly from ISO conditions as the ambient temperature increases on summer days. Turbine inlet cooling (TIC) air to ISO conditions (25°C) increases the performance of the turbine during the peak summer days, hours that coincide with critical peak electric loads.

If there is more than one turbine module, a district cooling system can serve all the modules; therefore, one DC station can serve all the turbine modules and benefit from a DC concept application.

Cooling Load Calculations

$$\dot{m} = \rho \times \dot{V} = \frac{P}{R \times T_{avg}} \times \dot{V}$$

$$T_{avg} = \frac{T_{indoor} + T_{outdoor}}{2}$$

$$Q_{cooling} = \dot{m} \times \left(h_{indoor} - h_{outdoor}\right)$$

where:

$\overset{\infty}{m}$ is the mass flow rate of air per turbine (kg/s)
V is the volumetric flow rate (m^3/s)
Q is the cooling rate (kW)
h is the enthalpy of air (kJ/kg)
P is the air pressure (pascal)
R is the specific gas constant (kJ/kg.°K)
T is the air temperature (°K)
T_{avg} is the average air temperature, indoor (downstream cooling coil) and outdoor (°K)

The cooling load curve graphs show the cooling load of the turbine inlet air during all the days in July, in addition to August 2014, and for June 15, July 15, and August 15, 2014.

The daily cooling load demand curve shows the daily curve, the average load, and the capacity of the TES tank. The block diagram shows the DC plant components and the three types of chillers used: direct gas-fired absorption—when there is a need to use natural gas, reject HRSG (heat recovery steam generator) exhaust absorption chillers—to use when excess reject heat is available, and centrifugal chillers—to use when excess electrical power is available. Chillers are operated according to the availability of energy.

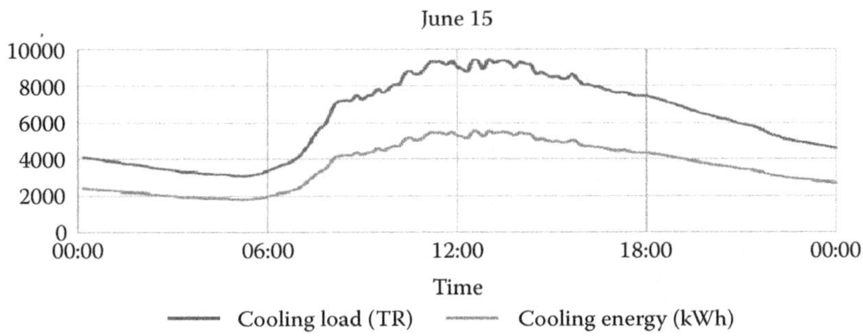

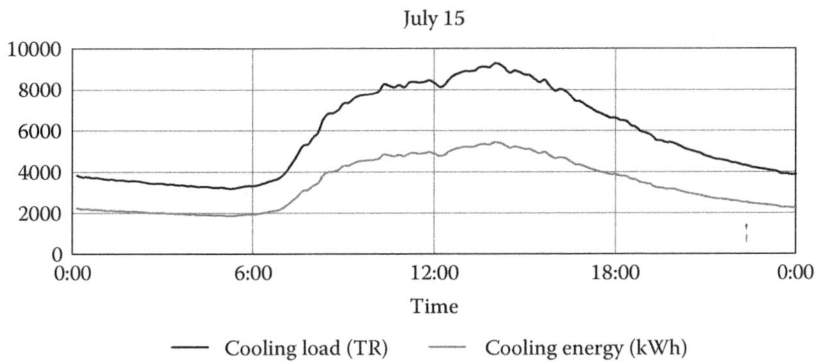

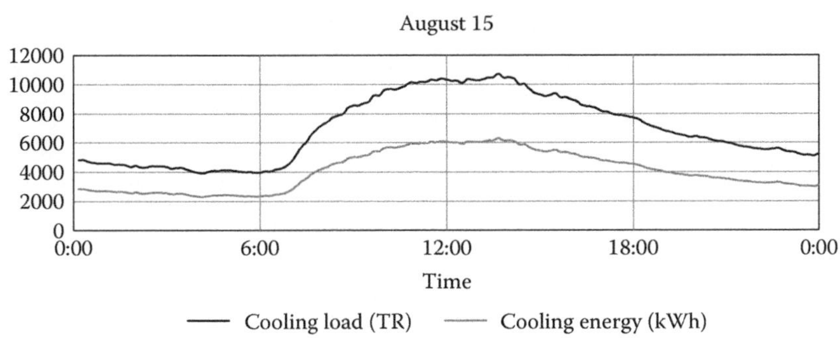

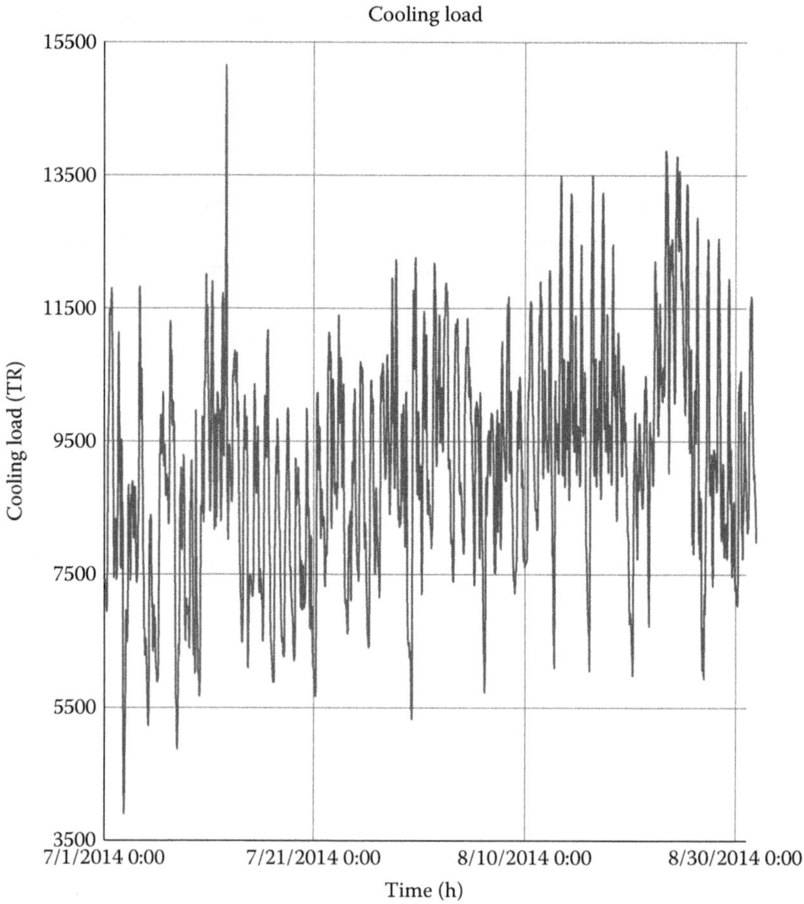
Cooling load
15500
13500
11500
9500
7500
5500
3500
Cooling load (TR)
7/1/2014 0:00
7/21/2014 0:00
8/10/2014 0:00
8/30/2014 0:00
Time (h)

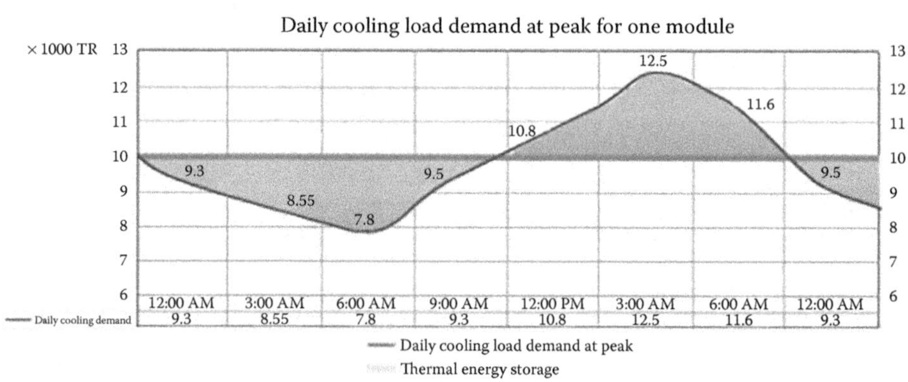
Daily cooling load demand at peak for one module
× 1000 TR
13
12
11
10
9
8
7
6
12.5
11.6
10.8
9.3
8.55
7.8
9.5
9.5
12:00 AM 3:00 AM 6:00 AM 9:00 AM 12:00 PM 3:00 AM 6:00 AM 12:00 AM
Daily cooling demand 9.3 8.55 7.8 9.3 10.8 12.5 11.6 9.3
Daily cooling load demand at peak
Thermal energy storage

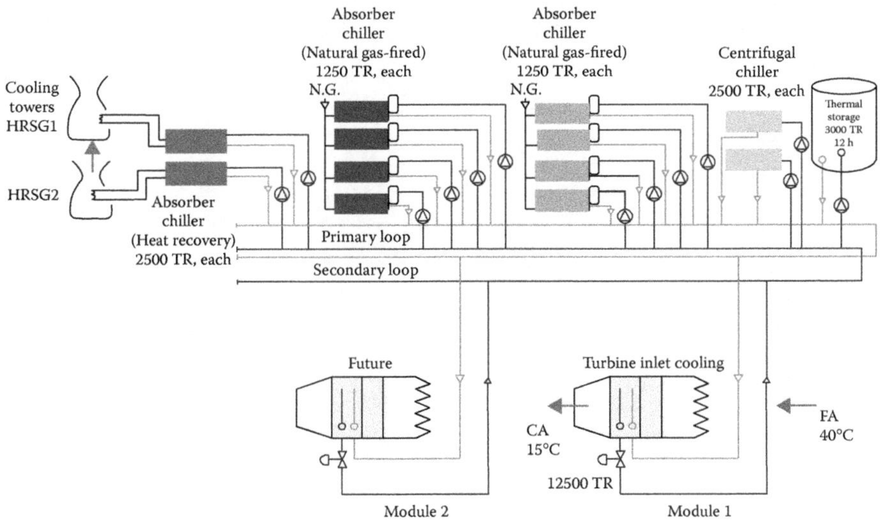
Absorber chiller (Natural gas-fired) 1250 TR, each
Absorber chiller (Natural gas-fired) 1250 TR, each
Centrifugal chiller 2500 TR, each
Cooling towers
HRSG1
HRSG2
N.G.
N.G.
Thermal storage 3000 TR 12 h
Absorber chiller (Heat recovery) 2500 TR, each
Primary loop
Secondary loop
Future
Turbine inlet cooling
CA 15°C
FA 40°C
12500 TR
Module 2
Module 1

Bibliography

1. El Sayed, T., Fayad, W., Monette, S.P., and Sarraf, G., 2012: *Unlocking the Potential of District Cooling: The Need for GCC Governments to Take Action.* Booz & Co. (now strategy&), New York.
2. UNEP, 2015: District Energy in Cities—Unlocking the Potential of Energy Efficiency and Renewable Energy.
3. Pierce, M.A., 2014: The Road to Lockport: Historical Background of District Heating and Cooling, ASHRAE Transactions: Symposia, for ASHRAE by Thomson Reuters (Scientific). www.techstreet.com.
4. Mornhed, G. and Casten, T.R., 2014: Innovations in District Heating and Cooling 1984–1994 and Their Economic Impact, ASHRAE Transactions: Symposia, for ASHRAE by Thomson Reuters (Scientific). www.techstreet.com.
5. Tredinnick, S., Wade, D., and Phetteplace, G., 2015: District Energy Enters the Twenty-First Century, *ASHRAE Journal.* www.ashrae.org.
6. Peng Kee, T., District Cooling as an Energy and Economically Efficient Urban Utility—Its Implementation at Marina Bay Business District in Singapore, Singapore District Cooling Pte Ltd., June 2010.
7. ASHRAE, 2011: *Applications Handbook*, Chapter 37—Owning and Operating Costs. www.ashrae.org.
8. ASHRAE, 2012: *HVAC Systems & Equipment Handbook*, Chapter 12—District Heating and Cooling. www.ashrae.org.
9. *Handbook for the International Treaties for the Protection of the Ozone Layer*, 5th edn, Climate Change 2007: The Physical Science Basis, IPCC, Contribution of Working Group 1 to 4th Assessment Report of the Intergovernmental Panel on Climate Change and UNEP, 2007.
10. IPCC/TEAP, 2005: Safe-Guarding the Ozone Layer and the Global Climate System, Issue Related to HFCS and Perfluorocarbons.
11. Refrigeration, Air-Conditioning, and Heat Pump Technical Option Committee (RTOC), Assessment Report, United Nations Environmental Programme (UNEP), 2014. Ozone.unep.org/sites/ozone/files/documents/RTOC-Assessment-Report-2014.pdf.
12. Herold, K.E., Radermacher, R., and Klein, S.A., 1996: *Absorption Chillers and Heat Pumps*, CRC Press.
13. Olama, M.A., 1980: Evaluation of some methanol-salt solution as absorption refrigeration solutions. PhD thesis, King's College, University of London, England.
14. Gosney, W.B., 2000: *Principles of Refrigeration*, Ashford Overland Services Ltd.
15. Nowakowski, G. and Busby, R.L., 2001: Advances in Natural Gas Cooling, *ASHRAE Journal.* www.ashrae.org.
16. Sulaiman, S.S. and Abdul Majid, M.A., 2011: *Gas District Cooling in Malaysia—Leveraging Abundant Cleaner Burning Natural Gas*, Lambert Academic Publishing.
17. Dorgan, C., Leight, S.P., and Dorgan, C.E., 1995: *ASHRAE Application Guide for Absorption Cooling/Refrigeration Using Recovered Heat.* www.ashrae.org.
18. G. Phetteplace et al., 2013: ASHRAE, *District Cooling Guide.* www.ashrae.org.

IMPORTANT GENERAL REFERENCES

19. *District Cooling Practice Guide*, International District Energy Association IDEA, 2008. www.districtenergy.org.
20. Frederiksen, S. and Werner, S., 2013: *District Heating*, Studenlitteratur.

Index

Note: Page numbers followed by f and t refer to figures and tables, respectively.

E

F

G

H

I

S

T

U

V

W